HOW TO READ AND CRITIQUE
A SCIENTIFIC RESEARCH ARTICLE

Notes to Guide Students Reading Primary Literature

(with Teaching Tips for Faculty members)

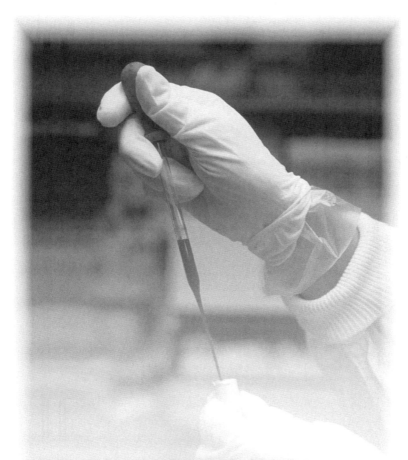

HOW TO READ AND CRITIQUE
A SCIENTIFIC RESEARCH ARTICLE

Notes to Guide Students Reading Primary Literature
(with Teaching Tips for Faculty members)

Yeong Foong May

National University of Singapore, Singapore

World Scientific

NEW JERSEY · LONDON · SINGAPORE · BEIJING · SHANGHAI · HONG KONG · TAIPEI · CHENNAI

Published by

World Scientific Publishing Co. Pte. Ltd.

5 Toh Tuck Link, Singapore 596224

USA office: 27 Warren Street, Suite 401-402, Hackensack, NJ 07601

UK office: 57 Shelton Street, Covent Garden, London WC2H 9HE

Library of Congress Cataloging-in-Publication Data
Yeong, Foong May, author.
    How to read and critique a scientific research article : notes to guide students reading primary literature (with teaching tips for faculty members) / Foong May Yeong, National University of Singapore, Singapore.
       pages cm
    Includes bibliographical references and index.
    ISBN-13: 978-9814579162 (pbk. : alk. paper)
    ISBN-10: 9814579165 (pbk. : alk. paper)
    1. Research--Methodology. 2. Peer review. 3. Critics. 4. Research--Study and teaching. I. Title.
    Q180.55.M4Y46   2014
    507.2--dc23

                                                       2013046111

British Library Cataloguing-in-Publication Data
A catalogue record for this book is available from the British Library.

Typeset by Stallion Press
Email: enquiries@stallionpress.com

Printed in Singapore

Contents

List of Figures

List of Tables

Preamble

The incorporation of scientific research articles as an essential part of the curriculum for undergraduate science students is becoming more prevalent, since more and more faculty members are finding out the value of exposing students to primary literature as part of their course-work (e.g. (Hoskins, Lopatto, & Stevens, 2011; Hoskins, Stevens, & Nehm, 2007; Kozeracki, Carey, Colicelli, & Levis-fitzgerald, 2006; Quitadamo & Kurtz, 2007)). Indeed, no longer is it sufficient for faculty to refer undergraduates to text-books as a means of acquiring information, since the emphasis, now more than ever, is on imparting thinking and learning skills to students at the university level rather than disseminating mere content knowledge (e.g. (Asai, 2011; Green, Hammer, & Star, 2009; Hurd, 1998; Wood, 2009)).

Given the explosion of information and knowledge in the field of Life Sciences, adapting primary literature as materials in course-work as part of active-learning seems to be more effective in improving scientific literacy among science undergraduates than the pure transmission of content knowledge using traditional text-books. In addition, students also will have to read research articles as part of undertaking laboratory research projects in higher undergraduate courses. This has been beneficial for preparing them for graduate school, though traditionally only students in later years of their undergraduate classes get to work with scientific articles.

As such, a good grasp of reading and analytical skills is needed for students to understand how their research project contributes to the field that they are working in. For students not intending to take on graduate studies, understanding the scientific process is useful as part of improving science literacy (Deboer, 2000) and this can be achieved in part through learning from the reading of primary scientific literature.

This book serves as a step-by-step guide for undergraduate students and faculty members dealing with scientific research articles as part of a module or research project. For the faculty member interested in using research articles to promote active-learning (Hoskins *et al.*, 2011, 2007; Kozeracki *et al.*, 2006; Quitadamo & Kurtz, 2007), I have provided here some ideas based on my own experience in using primary literature for my module. I use research articles in part or completely, when trying to engage students in data analysis as well as to introduce to them various aspect of scientific inquiry. By getting students to go through the process of reading a primary research article, they can gain more than mere knowledge, as such articles are not easy to read and require students to develop some level of competence in the field of research as well as critical thinking skills. Such skills that come gradually to students are applicable to them long after they leave university. There are specific suggestions I have for fellow faculty members which are highlighted in footnotes at the appropriate points. These are based on my personal experience in using primary literature in my attempts at reading- and writing-to-learn assignments using primary literature as the starting material (Yeong, 2013a, 2013b, 2013c). I have also attempted to introduce some aspects of scientific investigations so that students can better appreciate how scientific knowledge is constructed by the community of researchers. This will provide them a more complete overview of the scientific articles and some ideas about the nature of science.

For students encountering primary literature for the first time, it could be a rather daunting experience. However, with proper guidance, they can overcome the initial difficulties and become confident in dealing with scientific articles, as seen from previous studies (as cited above) and my own experience. Some courses do not provide step-by-step instructions as to how to begin reading an article due to time constraints but instead plunge directly into the activities related to reading scientific articles. Without proper guidance, most students are put off by the articles and would not read them voluntarily; they would only do so if reading articles is part of their project work or course requirement. Even then, they approach the articles with a certain level of apprehension and reluctance that can be hinder their learning experience.

To reduce the unnecessary stress linked to reading scientific articles, some guidance will go a long way. Using this guidebook, students will learn to highlight to look out for key points in a scientific article as they read their first research papers. An abbreviated form has previously been provided to my students in my assignments that students found rewarding. I elaborate in this guidebook essential information for students to note when reading a scientific article. For instance, the layout of a typical article, as well as the main aims in the methodology, data analysis and conclusions sections are explained, to give students some idea what a typical research article is about.

To help students acquire the capability to deal with the many aspects of a research article, pointers are provided to guide them to evaluate the possible strengths and weaknesses in the research article they are reading. This is so that they might learn how to judge for themselves, what the main findings of a research project are, the relevance of the findings relative to the current understanding in the field of the research, and how they might themselves be able to think about the methods to solve any biological questions of interest to them subsequently. On a more practical note, I also provide checklists at the end of each chapter that students could use to jot down key ideas they derived from reading as well as questions they might have as they read the articles. Such checklists can easily be adapted by students to suit their needs.

This guidebook, therefore, serves to help students ease into the task of dealing with a scientific article either as part of their required work or for finding out more about an area of interest on their own. Overall, as undergraduate students learn to overcome their reluctance or anxieties in reading scientific literature, they will better appreciate the process of scientific investigations and how knowledge is derived in science. This will improve the overall level of science literacy among Life Science graduates.

1

Introduction

The first time you are tasked to read a research article describing original research findings, try not to let the unfamiliarity of the article layout and the perception that research articles are incomprehensible deter you. First of all, do not feel that you have to understand the entire paper at one reading, as some of us take a few readings to appreciate the implications of the data presented in the entire article,[1] even after we have gone on to graduate school and beyond. Secondly, reading research articles improves with experience, so with more experience reading articles, you will get better and better as you go along. If it is your first attempt at reading a scientific research article, you might want to break the reading into several steps and go through the article systematically. It could help you to keep the key points in Table 1.1 in mind as you start. In that way, you will not feel too overwhelmed by the details, which you will be able deal with in a progressive manner subsequently (see following the chapters).

When you are reading scientific articles, it is crucial that you not merely read the articles and accept blindly all the points laid out in

---

[1] For faculty members dealing with first or second year undergraduates, I find that reiterating this point is important. Most undergraduates are capable of learning to read a research article even at their junior levels but are mostly intimidated by the idea of having to read a highly technical paper. A little reassurance will go a long way towards assuring students that research articles are not always specialised articles that only scientists can understand.

1

Table 1.1   Several Points Students could Bear in Mind as They Read Research Articles for the First Time

(I) What was the current state of knowledge in the field that the research work was described in the article? This information will be found in the **Introduction** section of the article.

(II) What was the idea tested in the article? This information will be found **at the end of the Introduction section** of the article.

(III) What were the key techniques used in the experiments described? This information will be found in the **Materials and Methods** section of the article.

(IV) What were the data obtained in the study? This information will be found in the **Results** section of the article.

(V) What were the conclusions drawn by the authors from their data? This information will be found in the **Discussion** section of the article.

the article. Rather, you should gradually develop skills needed to analyse the information and data critically. Hence, a good practice is to read through the article and as you do so, gradually gather the essential information as outlined in Table 1.1. After the initial reading to get at the basic information such as the background details and main objectives of the research work, you could re-examine the article more carefully in subsequent readings to judge for yourself if the data reported in the research article were reliable, valid and consistent with current understanding in the field. You can then further assess whether you agree with the authors' analysis of their data and conclusions in the Discussion section.

You might question why it is necessary for you to examine the article critically, if publishing in a scientific article usually means that experts in the field have already evaluated the work and deemed it of sufficient quality and novelty to warrant its publication. Well, as it turns out, even in scientific research where there is supposed to be objectivity in the way experiments are carried out, interpreted and reported, it is still run by scientists who might have their own preferred ideas and biases (scientists are after all humans). As such, articles are selected for publication not necessarily because the work represents an objective truth. Rather, articles are published because they have crossed the sometimes arbitrary threshold for novelty and technical requirements as accepted in the respective fields.

Moreover, there is also the possibility that more than one explanation could account for a biological phenomenon being studied and so, there might be issues that were not brought up in the article. So do not be too quick to decide that all that have been reported for a particular issue in an article means that the question that the researchers were working on is completely solved. You should also understand that not everything is included in a research article that is, on average, 5,000 words in length with a limit of 6 to 8 figures for data, depending on the journal. So, authors have to restrict the amount of data that they report. The reported work is therefore incomplete at times and far from perfect most times.

Hence, for you as a student, reading an article not only means learning from the authors in terms of the background information and ideas presented, but also, understanding that one should accept the ideas reported only if they are substantiated by the evidence in the form of experimental data. It is the skill to judge the strength of the data and how well the data support the authors' hypothesis that is pertinent to your development as a Life Science student.

How then do you "judge" the article? As with most things, it is easier if you broke your evaluation of the article down into several steps so as not to get overwhelmed by the sheer amount of information in the article. You could assess the article by individual sections, and for each section, assess its strengths and weaknesses (Fig. 1.1). Such a critical examination or critique of the research article is the basis for how other researchers assess the quality and impact of the work. Of course, you should know that to critique an article does not mean that you have to find only weaknesses in the article. Rather, it actually means that you have to decide, based on your prior knowledge or other evidence, whether or not what is stated in the article makes logical sense and is substantiated by experimental data.

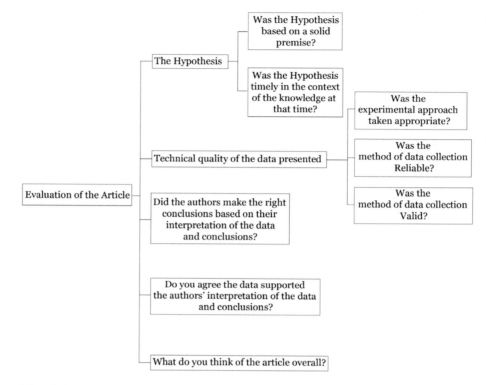

Fig. 1.1   Questions useful for evaluating the work described in the research article.

In relation to critiquing an article, it should be noted that for any point in the article that you consider a weakness, it is good to suggest ways to overcome the weakness, or propose alternatives to what the authors have stated. For instance, you might consider that the data reported in the article was not sufficiently convincing to support the authors' claims due to the lack of controls. You could then think about the types of controls needed to be included in theory that could improve the validity of the authors' assertions. This is constructive criticism[2] and should always be practiced, as it

_____

[2] From my own experience, some students tend to be a little too eager to criticise without a good basis. This stems perhaps from their inexperience and the misunderstanding that critique is equal to criticism. As such, getting them to think

is too easy to criticise a piece of work without providing any concrete suggestions. In other words, critique does not mean criticise.

Obviously when you are reading the article, you are not likely to write directly to the authors of the article with your suggestions. However, you could do this in the form of a thought experiment, or discuss with a classmate to train yourself to identify what in the article, is a weakness (for example, the experimental approach used), why it is a weakness and if you think something is amiss in the article, how to improve it (such as thinking of a better experimental strategy). Conversely, for an article that you think is good, you should also be able to justify why you consider it to be so. For example, the article could be good because of the authors' experimental design that allowed them to investigate a biological process using three different approaches. This could have led to a high level of reliability and validity in their findings that supported the conclusions the authors made.

In the following chapters, we will make use of the questions highlighted in Table 1.1 and Fig. 1.1 to guide you in making sense of a research article and reading it critically. This is not an exhaustive reference book detailing every tool you will need (such as how to use appropriate statistical tests or how to use various experimental techniques), but rather, to point out to you what you should look out for when reading a research article. Different examples from various fields of life science topics are used for illustrating specific points, so it is important that you understand the principles behind the issues I am discussing, as the ideas provided are transferable no matter which field you are interested in.

---

about why they think a particular issue in the article is a weakness and how they can suggest ways to improve the issue is an effective way for faculty members to demonstrate constructive criticism. Also, students should be reminded that they should state what is good about a particular issue as well, as critique does not mean looking at the negative aspects of the article.

2

How to Search for an Article

One of the initial steps to reading a research article is to find and retrieve the full article. In certain cases, your lecturer could provide you the article in its entirety. In other instances, you might only be provided the title and names of the authors of the article. You then have to look up the articles and download them at the appropriate journal or library websites. It is good to learn more about which search engines or databases are used routinely for finding and retrieving articles of interest.

When presented with the task of looking up articles, you have several key search engines to use to retrieve the articles of interest (Fig. 2.1). In such instances, the article information provided to you could be the authors' names or titles of articles. For Life Science and Biomedical articles, one key search engine is called the "Pubmed" hosted at the NCBI website (www.ncbi.nlm.nih.gov/pubmed). You can also use Google Scholar (http://scholar.google.com) or Web of Science or Science Direct (subscription required — usually accessible through your University library) to search for articles.

Pubmed is an example of a public online bibliographic database accessible to anyone with an Internet connection. This is a very powerful database that is maintained by the U.S. National Library of Medicine at the National Institutes of Health. The nature of the Pubmed is such that it not only serves as a bibliographic database for published works, but it is also a repository of the articles depending on the access that journals provide. For instance, in addition to providing bibliographic information of the articles of

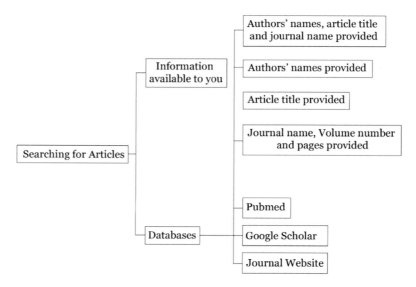

Fig. 2.1   Information needed to search for articles using specific databases.

interest to you, there are also hyperlinks out from each article listed to the full articles hosted at journal websites. In most cases, these journal websites require a subscription in order that the full article can be downloaded. If the library at your university subscribes to the specific journal in question, then you could easily download the article by going through your library service; most university libraries will allow matriculated students to access journals through the specific library links.

For research work funded by NIH and Welcome Trust, articles published in the various journals are made freely available at Pubmed. For certain journals, the authors pay a fee to the publisher so that everyone can access their articles in the journal. In other instances, journals might have adopted the Open Access format, while yet other journals make articles freely accessible a year after the initial date of publishing an article. How you can find out if a full article is available for download is to click on the link at Pubmed that would bring you to the journal that the article is published in.

At the journal website, you will find a link to a file. If it is an open-access file or if your institution has access to the article, you will be able to download the article in full. Otherwise, you will be prompted to sign in with a password. For articles with restricted access, you could find out if it is possible to request your library at your university to purchase a copy for you. The articles in journals can typically be viewed in a html format or downloaded in the pdf format. For certain articles, additional information might be provided in a Supplemental file that would provide detailed technical information on a specific protocol or technique. For those who want to find out more of how to make use of Pubmed, a quick-start guide is available at: (http://www.ncbi.nlm.nih.gov/books/NBK3827/#pubmedhelp.PubMed_Quick_Start), and a video tutorial (http://www.nlm.nih.gov/bsd/disted/pubmed.html) is available for first-time users.[3]

The other search engines such as Google Scholar or Web of Science or Science Direct function similarly to allow you to find your article of interest. Again, depending on the journals, the articles might or might not be available for downloading in full from these search engines. As mentioned above, you could alternatively directly look for your article at the journal website. This means going to the journal website and making use of the search function at the website to find your articles. Obviously this is a narrower search engine compared with Pubmed as it normally allows you to search for journal-specific articles. For the journals published by larger publishers, these search engines might be linked to the publisher's database that would allow you to search across various related journals under the publisher.

The easiest format to download is the pdf file that you could save and annotate off-line, for example, jotting down points of interest or queries you might have for your lecturer. The pdf file is easy to refer to and can be readily retrieved each time. On the other hand, the html format is also handy as you can click on links from the articles to other articles that have been cited. This will lead you

---

[3] It might be helpful for students to do a Pubmed search in class to provide students with hands on experience how to use it, especially to get from Pubmed to the specific journal website to retrieve articles in full.

to other articles that serve as background reading that will expand your understanding of the article of interest. However, annotation cannot be done in the html format so if you have the habit of jotting down reading notes, then the pdf format is more suitable.

The search for articles is not a difficult process and you could always gain access or get assistance to retrieving full articles through your library or lecturer. You might even request for the article from the authors directly if you are able to contact them. Once you have learned how to use the various search engines/databases, you could always look up other articles at these sites, knowing that these academic resources are a reliable source of articles as references.

3

Anatomy of a Typical
Scientific Article

After retrieving the article of interest, you could first scan through it to get a general overview of the article layout. Generally, you would see a typical organisation such as shown in Fig. 3.1.

If you look briefly through an article, you would normally find that the article title right is at the top of the first page, followed by the names of the authors with their affiliations (Fig. 3.2). The title normally gives a good description of what the research work is about. Usually the model organism used in the study is stated, together with the key idea in the article. The authors listed are the researchers who had contributed to the article in one way or another, such as conducting the experiments, designing the study, analysing the data or writing up the article. The order of the names normally reflects the extent of contribution of the authors to the research paper. The first author is the key researcher who has conducted most part of the experiments described, as well as data analysis and probably the writing of the article. The subsequent authors would also have contributed to various aspects of the research work, though possibly not to the extent as the first author. The last or senior author is typically the principal investigator who was in-charge of the project and had provided the ideas, funding and laboratory for the other authors conducting the research work. His or her contact details are provided for other researchers interested in corresponding with the authors concerning various aspects

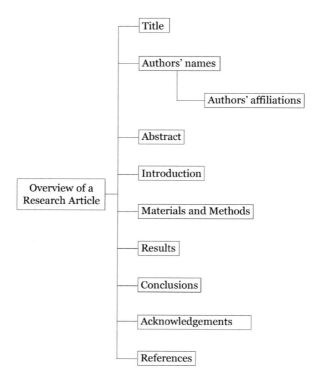

Fig. 3.1   A schema showing the key sections of a typical research article.

Fig. 3.2   An example of how an article looks like.

(© Chai *et al.* (2009) *Molecular Biology of the Cell*, ASM.)

of the work such as request for reagents used in the study and other detailed information about the data.

The authors' affiliations are an indication of where the authors are working or where the research work described in the article was performed. You might notice that the authors could be from different institutions. This is quite common for collaborative work where authors working at different institutions within or outside the country pool their resources or expertise to solve a common research question.

Before the start of the article proper is a short segment called the "Abstract" (Fig. 3.2) that is about 150 to 200 words in general (Chai, Teh, & Yeong, 2010). This is a summary of the key issue that the authors set out to study and includes the strategy taken in the study to tackle the research questions. The main findings obtained would also be presented. The Abstract further includes a statement or two about the significance of the data or conclusions from the study.

In some journals, other details shown at the beginning of the article include the name of the Monitoring Editor (Fig. 3.2). This would be the Editor who helped to coordinate the peer review process of the manuscript among the different reviewers. He also communicates the reviewers' comments and well as the final decision on whether the manuscript is of sufficient standards to be published in the particular journal.

After the Abstract you would see an "Introduction" section within which the authors will provide essential background information surrounding the research subject matter of interest. This is so as to lay down the premise of the research work. At the end of the Introduction, the authors would include a statement on the hypothesis they aim to test and some predictions of the possible outcome of their study.

The "Materials and Methods" section that typically follows the Introduction is an indispensable part of an article. In this section, what materials were used and how experiments were conducted in the research would be provided in a fair amount of details. The objective of the Materials and Methods section is to make available details of the experiments that would enable other researchers to repeat the experiments if needed. This could be for the purpose of verifying what the authors had reported as well as for other researchers needing the same or similar protocols for their own research.

The data of the scientific investigation will be presented in a section called "Results." In this section will be text descriptions of the results from the experiments conducted alongside the data collected from the experiments. The data that the authors have obtained in their experiments would be presented in the form of figures, tables or even videos (see more in the Results section). The implications of the data would be elaborated in a "Discussion" section. Here the authors try to provide an analysis of the data in context of the existing knowledge. They would attempt to explain how their data had explained or provided answers to their hypothesis.

Following these, there is a short section called "Acknowledgement" where the source of funding for the research projected reported in the article will be listed (Fig. 3.3). Further information such as the names of scientists or colleagues who provided reagents, materials and technical assistance or comments on the article would typically also be provided here. The last section in an article is the "References" section, which is essentially a list of articles cited by the authors (also known as a bibliography) in the article to support their work or are relevant to the study (more on this later).

Do note that this example of an article layout was taken from one of many journals and other journals would have similar organisation with slight differences. There could therefore be slight variations in the article layout as described above from other articles depending on the journals that the articles are published in. Different journals might organise the "Results" section before the "Materials and Methods" section and the citations and References could have alternative formats. Certain journals publish short articles where the "Results" and "Discussion" sections are combined as one. The details of how a research article should be presented in a journal is often stated at the specific journal's website. The instructions on how the articles are presented in a journal can normally be found at the journal website in a link called

"Instructions for Authors." This is meant for authors interested in submitting their work to the journal for consideration.

When the authors send in their manuscript to the journal for consideration to be published, they would have to follow the specific format. This would include the organisation of the text, the way to cite references, the format of the bibliography as well as the resolution of the figures and layout of tables. However, the format is mostly a stylistic issue that should not detract from the essential points in the article you are reading.

ACKNOWLEDGMENTS

We thank U. Surana, O. Cohen-Fix, S. Biggins, K. Hardwick, and M. Lisby for generous gifts of various strains and constructs. We also thank Cheen Fei Chin and Yuan Yuan Chew for technical assistance. We appreciate the insightful comments from the anonymous reviewers. We are grateful to Phuay-Yee Goh, Alice Tay, and Karen Crasta for comments on the manuscript. This work is supported by the BioMedical Research Council (Grant 04/1/21/19/325) and Ministry of Education (grant R183-000-168-112).

REFERENCES

Agarwal, R., and Cohen-Fix, O. (2002). Phosphorylation of the mitotic regulator Pds1/securin by Cdc28 is required for efficient nuclear localization of Esp1/separase. Genes Dev. 16, 1371–1382.

Alexandru, G., Zachariae, W., Schleiffer, A., and Nasmyth, K. (1999). Sister chromatid separation and chromosome re-duplication are regulated by different mechanisms in response to spindle damage. EMBO J. 18, 2707–2721.

Bardin, A. J., and Amon, A. (2001). Men and sin: what's the difference? Nat. Rev. Mol. Cell Biol. 2, 815–826.

Baskerville, C., Segal, M., and Reed, S. I. (2008). The protease activity of yeast separase (esp1) is required for anaphase spindle elongation independently of its role in cleavage of cohesin. Genetics 178, 2361–2372.

...

Fig. 3.3   Example of the acknowledgement and references sections. (© Chai *et al.* (2009) *Molecular Biology of the Cell*, ASM.)

4

A Brief Insight into How Scientific Articles Get Published in Journals

So what is the fundamental difference between publishing a scientific article in academic journals and publicising data on one's own webpage or in a self-published book? The central objective of publishing scientific research articles is to disseminate findings from experimental work so as to extend the knowledge in the field. Therefore, one critical difference is the process of peer-review that takes place when an article is initially sent to the journals for consideration to be published. This means that the work has to be properly evaluated by experts in the field who will judge whether the work has sufficient merit to warrant its acceptance to the journal for publication. Therefore, publishing one's work in a scientific journal involves stringent process that scientists in their respective communities are engaged in to ensure that the scientific findings reported formally that might influence and shape our understanding of biological processes are in fact based on proper scientific investigative methods.[4] As it stands, publication in scientific journals is part and parcel of academic pursuits and forms quite a substantial part of the process of conducting scientific research.

---

[4] It might be useful to explain this aspect of scientific literature to students so they understand that the process of establishing a body of scientific knowledge depends on scientific investigations and the evaluation of empirical data.

To understand a little more about research articles, it might be valuable to have some insight into how articles get published in journals. The process typically starts at the point in a research project when a substantial part of the project has accumulated significant data that could answer a specific question in the field. The authors who are keen to publish their research work in journals will have to write up their research work in a manuscript describing the details of the experimental data for evaluation by the journal. To write up the manuscript, the authors have to organise the manuscript into the various sections as described above (Fig. 3.1). The manuscript is then submitted at the journal website and the proper process of assessing the manuscript will be triggered. To get a more detailed idea about the submission guidelines and reviewing process, you could refer to the websites of two journals listed in Table 4.1.

Generally, manuscripts describing the research work will be judged on several criteria by reviewers who are specialists in the field. In such peer-reviewed journals where the manuscript is evaluated by researchers who are working in similar fields, one of the initial processes that will take place upon submission of the manuscript to the journal is an initial review by an editor who is working in a related field. The editor will ascertain if the work fits the scope and general standard of the journal. The editor might consult with other members of the editorial board to make a decision if needed.

If the work falls outside the scope of the journal or fails to meet the basic standards, the manuscript is returned to the authors so that they might submit their work elsewhere for consideration. If the work is within the scope of the journal and of sufficient quality,

Table 4.1   Samples of Author Guidelines and Criteria for Publishing in Journals

American Society for Cell Biology:
http://www.molbiolcell.org/site/misc/ifora.xhtml#C7
PLOS (Public Library of Science) journal PLOS Biology:
http://www.plosbiology.org/static/guidelines

the editor then assigns the manuscript to researchers in the field to act as reviewers for a more detailed assessment of the manuscript. These reviewers will evaluate the articles for the significance of the data to the field, the technical quality of the data, the strengths of the conclusions based on the data described and the novelty of the work. These are usually the main criteria for acceptance for publication in journals. The review process relies on 2 to 3 independent reviewers and can take between 1–3 months after the initial submission of the manuscript to the journal.

If the manuscript fails to pass the criteria as judged by the reviewers, the editor would reject the work and the authors can submit the work to other journals. If the manuscript passes the evaluation but has several areas needing correction, modifications or further experimentation as judged by the reviewers, the authors will be given a chance to address the issues raised by the reviewers and re-submit their work for a re-evaluation. There is usually a time period of 3 months for the authors to answer the reviewers' queries. This could mean performing further experiments if there were substantial issues to be fixed or changing the manuscript at the textual level. If the reviewers and editor are satisfied that the major issues are mostly addressed, the work is accepted for publication. If the reviewers are not satisfied with the modifications made, they can choose to reject the work. The summary of this process is shown in Fig. 4.1.

It should be noted that after a manuscript has been accepted for publication, other editorial processes would take place that are related more to transferring of copyrights from authors to the publishers of the journals as well as issues concerning the layout and editing of the text and figures. These are usually dealt with by the editorial team at the publisher and could take a few weeks before the article in its final format is published in print. Presently, with most of the journals going online, there are fewer and fewer journals printing entire issues of the journal as hardcopy, though authors could pay for their articles to be printed on paper. More often than not, pdf formats of articles are exchanged between researchers. Also, given the pressure to publish the latest research

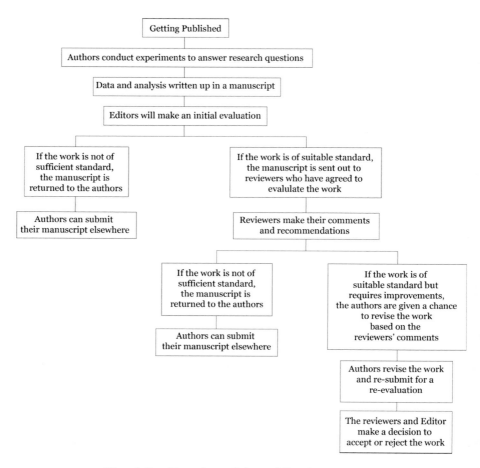

Fig. 4.1   Overview of the publication process.

findings, many journals are putting online the unedited versions of the articles while the editorial work is being done. This allows the work to be publicised as soon as possible.

There will of course be differences in the emphasis on certain criteria depending on the aims and scope of specific journals. Certain aspects such as novelty of the work and significance of impact to the field might be more relevant to some journals that aim to publish work at the forefront of the field. This does not mean that there is a compromise on the technical quality. Rather it means that in addition to high technical quality, the work must present data or ideas that are at the cutting-edge in the field and provides a leap in conceptual or technical advancement. In such journals, it would typically be stated in the website that novelty and mechanistic insights are needed so authors would understand that the requirements for publication in such journals are demanding.

In other journals, novelty and impact might not be as significant as technical quality of data. The articles published in such journals are also good quality work that might extend upon something previously known though not at a level that brings deep insights. The knowledge gained might only be incremental though it could still be of sufficient impact to warrant publication so researchers in the field can share the information. Again, most of such journals would state their guidelines for acceptance of publication. So it depends on the authors to decide which journals their work will best fit and then make a choice as to where to submit their manuscript for consideration.

In the scientific arena, publication of one's work is an essential aspect of sharing with others the findings that have been made that could advance the understanding of specific issues in the field of research. Needless to say, there is also a certain amount of competition between research groups working on similar topics. So to be able to publish findings before the competitor is one of the driving forces that motivate research groups to come up with innovative solutions to biological problems. However, at times, such competition might not be as productive and could lead to unhealthy rivalry between groups. In a positive light, one could view competition as working to improve the standards of the research in the field as each group strives to do the best science possible.

5

The Introduction Section: Background Information on the Topic of Research

The "Introduction" section is an essential part of an article where the authors lay down the background information related to the research project. In essence, the current state of the knowledge in the field is set out for the readers so they would appreciate it better when the authors try to explain how their research work will contribute new information to the field. Therefore, the Introduction can actually be a rich resource for students to learn more about the topic. For students keen to find out more than what is given in the Introduction section, there are the cited references that can provide further background information for the reader.[5] More significantly, the Introduction section also states the rationale and key question(s) that the researchers attempted to answer through their experiments (Fig. 5.1).

As a student, you might be apprehensive about whether you would possess sufficient knowledge to understand the Introduction. Normally, the Introduction is a relatively short section in the article due to word limits set by journals on the overall length of the article.

---

[5] From my experience, with a suitable choice of articles, faculty members can find articles where the Introduction material recapitulates the key concepts taught in class. When students read this, they feel good that they can understand this part of the article. This will help them overcome their initial hesitations to delve into primary research articles. By helping them to develop a sense of confidence in working with research articles, students can move on to reading more widely and more complicated articles in their later stages of undergraduate study.

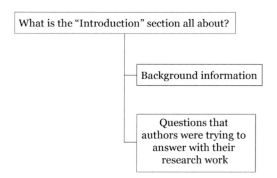

Fig. 5.1   The Introduction section lays down the essential information of the article as to why the authors embarked on their research project.

Unbelievable as it might seem, if you have gone through key lectures in your class on the topic related to the research article, you could most probably understand a good portion of the Introduction if the article has been specifically selected for your assignment. Just remember that the aim of an assignment is not to fail students but rather, to provide a task that should encourage the students to make use of what they have previously learnt. In assignments using primary research articles, chances are your lecturer wants to challenge you by stretching you further than simple made-up questions in a routine assignment. Thus, do not let the fact that you are about to read materials from a research article intimidate you, but rather, take it as an opportunity to discover for yourself, what it is that scientists actually do in a research laboratory, using the knowledge you have acquired in class to read about their work in the article. In the event that you have doubts about the information provided in the Introduction, do clarify with your lecturer so that you have a clear picture of what the authors are trying to study.

As an example of an article that can be understood by students in Cell Biology (second-year class), I have selected the article by Hating, Karlsson, Clute, Jackman & Pines (1998) on the Mitosis Promoting Factor. For students taking cell biology at the introductory level, the Introduction section of this article is no more demanding than what can be found in chapters on mitosis in a typical molecular and cell biology textbook and will not pose significant difficulties for undergraduate students who have attended lectures on cell division covering key ideas on mitosis. In fact, from informal surveys I have conducted with my students who took my course and have used such articles, we found that the students could understand these scientific articles more readily than expected. What is important is that students should always try to clarify any doubts they may have and not let any questions they have accumulate.

Another salient feature of a research article that students should note is the style of writing. As mentioned before, the background information related to their research topic of interest is laid down by the authors in the Introduction section. It is important to pay attention to how the authors cite[6] or make references to prior published work to support the statements they make.

So what is the significance of such "cited articles" or citations? Why do authors of scientific articles or other academic writing go through the trouble of citing previously published articles? This does not seem like a natural way to write! These cited works are generally other research articles related to the work described in the article. As shown in Table 5.1 such citations include the authors' surnames and year (the latter within parenthesis). Alternatively, citations could be given as superscript numerals in the text. The format used depends on the style adopted by the journal. However, the style actually has no influence on the quality of the article. All

---

[6] The use of cited work to support a statement in the Introduction section is an essential part of teaching students to make claims with support of empirical data. As traditional lectures are mostly about the lecturer stating facts that students accept, students might not be aware that the facts or knowledge told to them actually have been established through experimental data. By introducing the notion that scientific writing in research papers makes use of substantiated statements, students would be prompted to examine their statements more closely when they subsequently write a scientific essay or thesis, as they have to decide how to make use of the data to support their claims. More importantly, this idea should also impact their view of where knowledge comes from, as the sources of scientific knowledge in the form of cited work are available to them to look up in the reference list at the end of the article.

Table 5.1   An Example of Highlighting Citations within a Scientific Article

Excerpt:

citation ⟶ Chitin synthase 2 (Chs2) lays down the septum at the mother–daughter neck (Walther and Wendland, 2003; Cabib, 2004) as the plasma membrane is pulled inward during the constriction of the actomyosin ring in cytokinesis (Schmidt *et al.*, 2002; VerPlank and Li, 2005). *CHS2* is expressed in metaphase (Pammer *et al.*, 1992; Choi *et al.*, 1994; Spellman *et al.*, 1998) but is retained in the endoplasmic reticulum (ER; Zhang *et al.*, 2006) by the phosphorylation of a cluster of N-terminal Ser-Pro sites by the yeast cyclin-dependent kinase (Cdk1) Cdc28 (Martinez-Rucobo *et al.*, 2009; Teh *et al.*, 2009).

In the "Reference" section at the end of the article, you will see a list of cited articles. For instance:

References (abbreviated list shown)

...

Choi WJ, Santos B, Duran A, Cabib E (1994). Are yeast chitin synthases regulated at the transcriptional or the posttranslational level? *Mol Cell Biol* **14**: 7685–7694.

...

Martinez-Rucobo FW, Eckhardt-Strelau L, Terwisscha van Scheltinga AC (2009). Yeast chitin synthase 2 activity is modulated by proteolysis and phosphorylation. *Biochem J* **417**: 547–554.

...

Pammer M, Briza P, Ellinger A, Schuster T, Stucka R, Feldmann H, Breitenbach M (1992). DIT101 (CSD2, CAL1), a cell cycle-regulated yeast gene required for synthesis of chitin in cell walls and chitosan in spore walls. *Yeast* **8**: 1089–1099.

...

Spellman PT, Sherlock G, Zhang MQ, Iyer VR, Anders K, Eisen MB, Brown PO, Botstein D, Futcher B (1998). Comprehensive identification of cell cycle-regulated genes of the yeast *Saccharomyces cerevisiae* by microarray hybridization. *Mol Biol Cell* **9**: 3273–3297.

...

Teh EM, Chai CC, Yeong FM (2009). Retention of Chs2p in the ER requires N-terminal CDK1-phosphorylation sites. *Cell Cycle* **8**: 2964–2974.

...

(Continued)

Table 5.1   (*Continued*)

VerPlank L, Li R (2005). Cell cycle-regulated trafficking of Chs2 controls actomyosin ring stability during cytokinesis. *Mol Biol Cell* **16**: 2529–2543.

…

Walther A, Wendland J (2003). Septation and cytokinesis in fungi. *Fungal Genet Biol* **40**: 187–196.

…

Zhang G, Kashimshetty R, Ng KE, Tan HB, Yeong FM (2006). Exit from mitosis triggers Chs2p transport from the endoplasmic reticulum to mother-daughter neck via the secretory pathway in budding yeast. *J Cell Biol* **174**: 207–220.

(Adapted from (Chin *et al.*, 2012); © 2012 Chin *et al.* This article is distributed by The American Society for Cell Biology under license from the author(s).)

the citations in an article are normally listed at the end under a "References" or Bibliography section (for example, see Fig. 3.3) so readers can look them up for more details.

To explain this further, the Introduction section of the article by Chin, Bennett, Ma, Hall, & Yeong (2012), reporting on the regulation of an enzyme called chitin synthase 2 (Chs2p) that lays down a chitin wall between dividing yeast cells in the article is used. The full title of the article is: Chin, C. F., Bennett, A. M., Ma, W. K., Hall, M. C., & Yeong, F. M. (2012). Dependence of Chs2 ER export on dephosphorylation by cytoplasmic Cdc14 ensures that septum formation follows mitosis. *Molecular Biology of the Cell* **23**(1), 45–58. doi:10.1091/mbc.E11-05-0434.

In the foregoing passage, the characteristics of the enzyme called chitin synthase 2 (Chs2p) is being introduced. In the first statement on the role of Chs2p in laying down the septum at the neck of dividing yeast cells, reference was made to two other articles in parenthesis (i.e. "Walther and Wendland, 2003 and Cabib, 2004"). If you looked up the cited works in the References list at the end of the article, you would find the details of these two articles by the respective authors. These two are review articles that have surveyed the work on Chs2p and the authors in the article were using these review articles as authorities on the previous experimental work that provided data on Chs2p at the neck. Further on, the statement on *CHS2* being expressed in metaphase is attributed to the work by the research groups Pammer *et al.* (1992); Choi *et al.* (1994) and Spellman *et al.* (1998). Again, going to the reference list as provided above, you will be able to find in detail the articles by these groups that showed data on the detection of Chs2p during metaphase through their experimental work.

This use of prior work to substantiate claims is an essential feature of scientific writing. That is to say, scientists make statements of claims using empirical data published previously to support their points of view or statements of facts. This is a fundamental skill to bear in mind when writing a scientific report or making scientific arguments in an essay.[7]

Points to Look Out for in the Introduction

The Introduction section could allow you to review what you have learnt in class if the article has been chosen specifically to cater to your learning needs as part of your module or perhaps a mini-research project. In addition, new information may be garnered that could add to your learning about the topic of interest. To appreciate the significance of the findings described subsequently in the article, pertinent questions as outlined in Fig. 5.2 have been included to guide you in your reading and appreciation of scientific articles.

---

[7] For writing assignments, I sometimes ask students to read only a single article for their first essay-writing assignment. In this case, I would remind students that they have to use supporting data by citing the specific figure number of a piece of data they are referring to in order to substantiate their statements. Such a style of writing is not intuitive for students and examples could be provided to further reinforce this notion. With practice, students can learn to write in an academic manner that would be useful for them in later years of their studies. More importantly, learning to substantiate their claims with evidence is a very relevant skill for them to appreciate the nature of scientific knowledge.

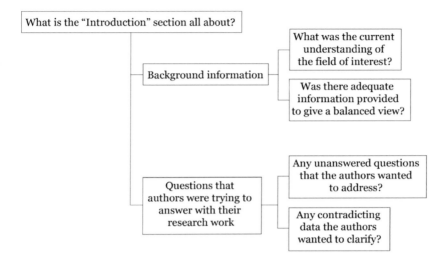

Fig. 5.2 The Introduction can provide useful information for student's initial reading of the article.

In summary, the Introduction section of an article presents the existing ideas in the field highlighting the current state of knowledge in the field. This will provide a proper context in which readers could consider the research questions that the authors are attempting to answer. That is to say, if the authors have set the background information on their research topic clearly, readers are better able to understand what the reasons are that the authors embarked on their study, be it a need of the authors to address an issue that is not clear in the field or a need to work on unanswered questions about their topic of interest that required further studies, given the state of understanding in the field at the time the authors did their work. For instance, was there a previously unknown detail about a protein structure that the authors were interested in, or was there a gap in the understanding of a particular part of a biochemical pathway that the authors wanted to fill?

Furthermore, it could be that the authors had contradicting observations and interpretations (see next section) in the research field that could form the basis for the study in the article. In such an instance, the authors could sometimes propose a view challenging that of the current view in their research field. The authors could be proposing to re-examine such issues perhaps using new tools and technologies not previously available so as to clarify confusion in the field. So as you read the Introduction, do keep these questions in mind and try to understand why it is that the authors embarked on their research work. This would contribute towards your appreciation of the article when you finally finish reading it.

For more advanced students who have read a bit more broadly, a further question to consider might be whether the authors have provided a balanced perspective of opposing views in the research field. For instance, did the authors cite representative work from researchers holding opposite views of the field? Objectivity is a foundation of scientific research so you should be on the alert if the authors only presented one perspective even though there might be others.

A simple way to keep track of the key ideas presented in the beginning section is to jot down quick notes in a checklist[8] (see Table 5.2) that serves to remind yourself, as you move along the article, what the knowledge is surrounding the experiments at the point when the article was published. More importantly, you should take note of issues you are unsure about so that you can clarify them before you delve deeper into the article; otherwise, you might accumulate too many questions that would hinder your reading of the article. This will help your overall grasp of the article as you progress in your reading.

A simple way to make good use of the Table 5.2 is to keep a soft-copy of the checklist and update it as you read the article and clarify issues that you might have. Over the course of time, the list will be filled with pointers and answers that you have accumulated

---

[8] Such a checklist or list of questions can be used as part of a class activity to get students to keep a log of their efforts in reading the research article. Or students could simply be encouraged to develop a habit of making notes of what they have read so as to keep track of what they have and have not understood when reading a research paper.

Table 5.2   Checklist to Help Students Read the Introduction

Questions	Remarks
What were the main points in the Introduction?	1. 2.
What did you not understand in the Introduction?	1. 2.
How did you clarify your doubts?	Discussion with classmates? Did you resolve your doubts? Clarification with lecturer? Did you resolve your doubts?
What questions did the authors highlight that were unresolved in the research field that they wanted to work on? (This normally forms the basis for undertaking the research that they described in the article.)	1. 2.

and you will see how much you have progressed in terms of your understanding. This can be done for the first few articles until you become more familiar with reading that you might not need it anymore. Alternatively, this can be a tool to carry on using if you deem it useful and your notes can be shared with classmates during discussions.

6

More on the Introduction Section: Hypothesis or Question that the Authors were Investigating

As mentioned earlier (Fig. 5.1), the authors would lay down the research question in the Introduction section. This is typically towards the end of the Introduction, where the authors would propose testing of a specific idea or hypothesis so as to establish new information or novel perspectives in their research field. Broadly-speaking, the term "hypothesis" would entail a premise on which one builds an idea to explain the underlying basis of a particular phenomenon or observation that one wants to test. As there are different research approaches, so different types of work are published, for instance, articles detailing work that aim to resolve outstanding issues or other types of work that challenge existing views. Generally, there are two distinct approaches that research can be conducted. One type would be a "question-driven" and the other "discovery science-driven" research[9] (Fig. 6.1).

---

[9]The distinction between research articles in terms of whether they describe "question-driven" or "discovery science-driven" research is not necessarily clear. However, it might be a good teaching point to highlight the notions of what is question-driven research in general as compared with the more discovery-driven types of work. This will allow students to appreciate the motivations underpinning research work.

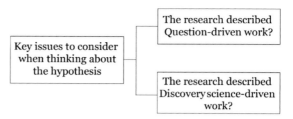

Fig. 6.1   Two broad categories of research articles.

In "question-driven" research, the authors had an idea that they wanted to test through conducting their research work. For instance, the authors could propose to test ideas that could provide mechanistic insights into a particular process by probing deeper into how 2 previously identified proteins might interact with each other. Also, the authors could have preliminary observations that might be contradictory[10] to the prevailing ideas in the field and therefore propose alternative hypothesis to investigate their preliminary findings. The authors could also seek to dispute existing ideas using newly invented technologies not previously available so as to establish a new theory.

In "discovery science-driven" research, it is typical that a more general question is asked and then data would be uncovered on a large-scale manner and then analysed for specific patterns that could provide ideas to explain certain phenomenon. Such large-scale projects are usually driven by advances in technological capabilities. The Human Genome sequencing project is a good example of such "discovery science-driven" research supported by technological advancements. Other types of "discovery science-driven" work include comparative studies to carry out a line of investigations

---

[10]For faculty members interested in using contradicting ideas to highlight the nature of scientific investigations, articles that challenge the prevailing views about a particular topic are useful for introducing "real-life" examples of how researchers can disagree on ideas. Moreover, such articles can be useful for engaging students in evident-based arguments on a particular scientific idea. This is an important part of student learning that is normally not easy to encourage in a traditional classroom setting if the lecturer was merely stating the accepted views in the field.

that would extend the understanding in the field, such as examining a known biological process but in an organism not previously characterised. In other cases, it could be to survey the gene expression of a panel of tumour suppressor genes in a particular type of cancer samples to determine if there were patterns amongst the clinical samples from patients. If some pattern of gene expression emerges from the analysis, the information could be useful at the level of diagnosis to predict the outcome of the patients or the use of a particular treatment regime that might be beneficial to a certain class of patients with specific patterns of gene expression.

It is obviously not always easy to categorise research work unambiguously into "question-driven" and "discovery science-driven" as it might be possible to find research articles describing work comprising both approaches.[11] In terms of the approach towards Life Science research such as in the molecular biology field, traditionally, researchers based their studies on one-gene one-function approach to studying a particular biological process. With the advent of precise robotics and computational capabilities, large-scale or genome-wide experiments can be carried out to examine thousands of genes and gene products in a high-throughput manner. So one can ask a broad question and harness the new technologies to discover more genes and gene functions than one-gene one-function approach. There is also no necessity to categorise the research described as you read various articles unless your have specific instructions in your assignment to do so. Nonetheless, it is valuable to identify the aims and overall strategy of the research described in the article, as this will shape how you could understand why the research is of importance. This might also influence

---

[11] Other than the fact that it is not an easy task to categorise the types of research articles, it should also be pointed out to students that neither approach is the superior one, as both the question-driven and the discovery-science strategies have their strengths and weaknesses. When used in combination, the limitations of either method can sometimes be overcome. However, the decision to use one or the other approach depends on the research project and resources and as such, students should recognise that research articles are not usually reports of completed or perfect research work.

how you could assess the work in light of the underlying research questions.

Issues to Consider Regarding the Hypothesis or Question

There are several areas (Fig. 6.2) you should bear in mind when reading the Introduction that could help provide some perspectives on the article. Breaking down your analysis into separate criteria for evaluation will help you organise your thoughts better. Also, with a systematic approach, you will not feel overwhelmed by the prospect of having to read a scientific article completely for the first time.

a. *Premise of the work*

To begin evaluating the article, you could start with the premise of the work, i.e. the hypothesis or research question. Based on the Introduction as well as your knowledge of the field of research, do you think the proposed work was relevant in the context of the knowledge at that time? That is to say, was their research question relevant in addressing a particular issue not currently understood? Or are the authors trying to open up a new field using new technologies? Could the authors be trying to put up alternative ideas to what is currently known in the field? These are the usual types of

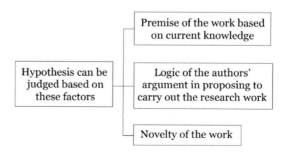

Fig. 6.2   The authors' hypothesis can be broken down into several areas.

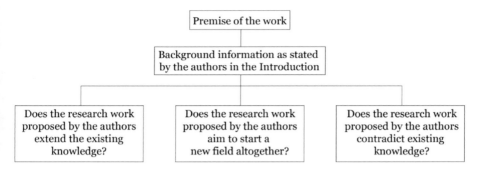

Fig. 6.3 Thinking about the premise of the work.

questions that researchers ask when they are conducting their research work, so it would be valuable for you to use these as pointers as a guide you understanding the authors' hypothesis (Fig. 6.3).

To answer such questions would require you to know the field of research, which might not be easy if you were reading scientific articles for the first time. Nonetheless, you could form some initial ideas about the topic from the background information laid down in the Introduction and then read more articles around the topic, as you get more familiar with scientific literature. You could also work with your classmates by sharing reading topics outside the article, followed by discussions among yourselves.[12] The level of work you could do depends on the task you have and the time frame provided for your assignment; therefore, plan your time to determine the extent to which you need to read outside of the materials required of you.

For students keen to find out more, an easy way to get to know the field of research is to read up recent review articles the topic of the research article. Review articles are overviews of the current

---

[12]This could be an opportunity for a peer-learning activity if time permits. Students could be assigned different review articles to read around the topic related to the research article. A discussion session or online forum could be organised for students to exchange some ideas about the background information. Such activities are beneficial to broadening students' perspectives as they get to read outside of textbooks.

knowledge written by established researchers in the field. In such reviews, the key understanding of the topic is presented, with supporting evidence from primary literature. By getting a grasp of the consensus views in the field, one will be able to make a simple judgment as to whether the proposed study has indeed identified gaps in knowledge or if new avenues are explored.

b. *Logic of the work*

Another related question to bear in mind about the proposed hypothesis is whether it has a sound basis. This would require you to judge at least the logic of the authors' justifications for carrying out the research work that they described in the article. Related to this, you could also think about whether the authors have to make any assumptions in their hypothesis. If they did, are the assumptions substantiated by supporting evidence?

For example, supposing that in the Introduction section of an article, say, on coral reproduction, the authors might have elaborated on information relating to a newly-identified species of corals with respect to asexual reproduction that occurs via the budding off of small polyps from a main coral head. They also mention that the temperature of the environment affects the rate of asexual reproduction in this species. To understand the role of temperature in the asexual reproduction process, the authors could propose to study whether the temperature affected the rate of budding of polyps from the main coral head or the survival of the polyps that had budded off. These are two possible testable ideas that could explain how temperature fluctuations could lead to a change in the rate of asexual reproduction of the corals. These ideas are more logical than if the authors were to propose to study how temperature might affect intake of nutrients by the corals as a measure of asexual reproduction rate. In this case, they were making the assumption that intake of nutrients is linked to asexual reproduction. When such an assumption is made, do they have evidence to substantiate this claim? If there is none, are they also proposing to show how intake of nutrients is linked to asexual reproduction? If they are not, then their assumption lacks a strong basis and thus would not hold.

In another example, the authors might state that a particular type of breast cancer is closely-correlated with the loss-of-function of a gene X. If gene X were not well-characterised, the authors could propose to study the expression of gene X in relation to how it is normally regulated in untransformed cell lines that originated from breast tissues. They could also study how such a process could be disrupted in breast cancer cell lines to try and provide some molecular basis for the down-regulation of gene X in cancer samples. Would you consider that a reasonable approach to study the expression of gene X following from the observations? In a sense, that could be a typical approach to investigating the basic molecular biology of how gene X is controlled under normal circumstances (using the untransformed breast cell lines as the model system). Using breast cancer cell lines to further study how gene X expression could be lost also falls in the general strategy, although it could be argued that the patients' samples might not necessarily be the same as the cancer cell lines, though the cancer cell lines are accepted models for the actual cancer tissues. In such a case, the assumption is that the findings from cancer cell lines could be extrapolated to what is happening in human bodies leading to cancer development. Such assumptions are generally made due to the absence of a good way to study cancer formation and progression directly in a human body (other than ethical constraints restricting certain types of studies using humans as subjects for experimentation).

c. *Novelty of the work*

The novelty of the research question could also be an issue for you to consider in relation to the hypothesis. To think about the novelty factor of the article means that you need to consider if the proposed work will provide new information about the field. Of course, this might not be an easy criterion for you to judge, as it also relates to how familiar you are with the field of the research. As you read a research article for the first time, this is sometimes a little too challenging.[13] In your first article, you could simply consider how the proposed research might provide novel information or new understanding compared to what the authors have provided in the Introduction.

Furthermore, the extent of the novelty will vary from article to article, with certain articles with in-depth studies providing more new findings than others. Normally, most journals would state in their website that this is a key criterion for acceptance in the journal. This depends on the journals, as each will set the agenda for the quality of research articles. Nonetheless, as a novice in reading articles, you might still find it worthwhile to compare the findings in the articles to the background information provided in the Introduction for practice.

Obviously this is not an easy thing to do as you read your first research article, since you would need to know the field rather well. Nonetheless, as mentioned in the chapter on Introduction, you could get to know a little more of the field by reading one or two review articles. In addition, you could consider how much more information had the results in the article extended the current understanding in the field even by just reading the particular article.

Alternatively, you could form a discussion group among classmates to share some ideas about whether you think the work will be informative and to what extent. In some cases, the novelty factor of the research article could be the use of a new innovative technology to study an established biological question. Or you could sound out your lecturer. In all, it is at least useful for you to appreciate the idea that publishing a research article is meant for the

---

[13] In addition to setting up peer-learning activity (see Footnote 12), faculty members can help students on this aspect perhaps by providing a short synopsis of what is presently known in the field. They can then ask students to read the article and then compare against the synopsis. For more mature students, they could be directed to review articles that will provide sufficient background information on the topic.

researchers to report on something that would be of interest to other researchers in the field or scientists in general, because there is something not previously known.

There is also the issue of relevance of the research work to the field of interest or perhaps to a part or whole of society, in terms of providing a novel perspective on a scientific issue. This can sometimes be a problem in the realm of scientific research, as certain topics might be ahead of their time and so appear to be irrelevant or not as popular to the scientific community at the time. Such pieces of work likely get relegated to small obscure journals although the research might be technically good. They might eventually become relevant when the topic matures sometime down the road due to ideas coming together that draw interest to it. At that point, articles describing work in this area may become more acceptable for publications in higher-profile journals. So we see topics coming into and out of "fashion" so to speak.[14]

Understanding the aims and hypothesis of an research article will help you form some initial ideas about the directions of the paper at the start of your reading. The hypothesis will drive the Methods and consequently, the Results obtained from the research work. The Results should allow the authors to form the basis to support or reject the hypothesis proposed at the beginning of the article. It also allows the reader such as yourself to judge if you would agree with the authors' conclusions based on their arguments using the Results of their study.[15] As such, after you have

---

[14] Depending on your module objectives, it might be a valuable part of students' scientific training to provide some insights into the social aspect of the scientific enterprise. This exposure will allow students to think about issues relating to objectivity in a way certain science topics gain or lose favour among the scientific community, depending on the prevailing ideas. This has influence on the way society thinks and/or allocates funds to specific areas of scientific research.

[15] As part of getting students to understand the scientific world and to encourage them to think a little more critically, I do sometimes highlight this point so that the students would not automatically accept all that they read in textbooks or articles. This is to me an essential part of keeping the teaching authentic by making students evaluate carefully what it is that they choose to accept as objective truths in science.

Table 6.1   Checklist for Examining the Hypothesis

Questions	Remarks
Was the background information provided in the Introduction balanced?	
Was the project based on a solid premise?	
What was the aim of the work? • Extending existing knowledge • Describing a new technique • Challenging existing knowledge	
How would you judge the novelty factor of the hypothesis? • Based on the reading of the Introduction • Following some reading of reviews around the topic (if you have time) • After discussion with classmates • In consultation with your lecturer	

read the entire article, it is good practice for you to come back to the Introduction to evaluate the article in its entirety. An overview of the issues to consider in respect of the hypothesis of an article is shown in Table 6.1.

7

The Materials and Methods Section: Reagents and Techniques Used in the Study

A general rule about the Materials and Methods section is that the information provided in this section should allow other researchers to repeat the experiments described in the article.[16] This is essential, as it makes available, the different methods, especially the new ones, to interested researchers to conduct the same or related experiments. Interestingly, long ago it used to be that the editors of journals repeated the experiments described in research articles so as to ensure the reliability of the data reported by the authors.

Nowadays, such a practice of by editors no long takes place due to the large number of manuscripts submitted to journals. Therefore, including the Materials and Methods section is more a critical part of sharing information among scientists that could help improve the overall techniques and protocols used for researchers in the field (Fig. 7.1). Also, the Materials and Methods used for

---

[16] This could be a good opportunity to impress upon students the value of good record keeping in the students' own practical work and research projects, as may be the case. More often than not, I found that students doing research projects do not sufficiently appreciate the need to record their protocols (and data) properly because they might not understand why it is an essential skill in the field of research. Also, the practice of sharing information among scientists is an important lesson that students should learn about as this is the foundation upon which advances in scientific enquiry is built.

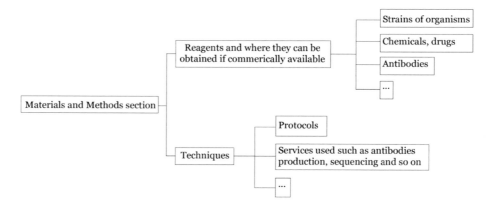

Fig. 7.1   Information provided by the Materials and Methods section.

data collection form an essential part of the paper, as the results are dependent on the types of materials and techniques used in the experiments. Hence, the details provided will enable other researchers to evaluate the work.

Information that would be included in the Materials and Methods section comprise details such as where chemicals and reagents are bought, how plasmids and constructs are built and the sources of organisms used such as cell lines or mouse strains. This information is useful for other researchers to obtain or purchase the reagents or materials needed for their studies.

When you examine the Materials and Methods section, you will find all the different tools and techniques[17] that the authors used in their experiments. For instance, the tools could include equipment used for specific experiments, together with the reagents and experimental organisms used. The techniques would refer to particular methods and protocols that were followed in the experiments. The information can sometimes be elaborated in full for newly established protocols. There could also be short descriptions

---

[17] Teaching students about individual techniques relevant to specific experiments might require dedicated explanations, depending on module requirements of faculty members. When I have reading assignments, I will try to incorporate techniques relevant to understanding research articles in my module teaching so that students are familiar with certain technical terms and applications of different techniques.

of previously established protocols. For protocols previously reported, it is sufficient to cite the original articles where it was first described. The extent to which details are provided depends on the journals, with some journals providing more space and others less so. In some cases, the information on certain methods is allocated to a section known as the Supplemental materials. This is a more recent practice due to the increase in the information and data that are now published in a typical research article in the Life Science fields. These details are usually presented in sub-sections within the Materials and Methods section (Fig. 7.2).

Information such as the ethical handling of organisms (for example, transgenic mice) or how patient confidentiality issues in drug studies were maintained; whether institutional ethics reviews were conducted for such studies, or if exemptions were granted, would be explicitly stated. Also, readers are made aware of whether ethical standards were maintained for the research work.

In certain instances, if detailed information was too long, the information could be provided in Supplemental sections of the article. The Supplemental sections could be made available only online in certain journals in html format. In other journals, this section could be downloaded together with the main article in pdf format. Sometimes the Supplemental section includes data that are considered not critical enough to be included in the main Results section.

Normally if other researchers would like to make use of certain materials such as specific strains of organisms that are not commercially available, they could request them from the authors. Such exchange of materials not commercially available or information on details of protocols is a key criterion that is part of the scientific community's effort to promote sharing of materials and knowledge. In fact, journals usually state as part of the condition for the publication of the research articles the need to distribute reagents freely if other researchers request for them.

Depending on your assignment, it might be interesting to read about the technical aspects described in the Materials and Methods section if you had come across the techniques during the course of

MATERIALS AND METHODS
Yeast culture reagents

Supplemental Table S1 lists the strains used in this study. Cells were routinely grown in yeast extract peptone (YP) or selective medium supplemented with 2% dextrose at 24°C. For experiments requiring Gal induction, cells were grown in YP supplemented with 2% raffinose (Raff), followed by addition of Gal to a final concentration of 2% unless otherwise stated. Each experiment was performed at least three times, and 100 cells were counted for each time point unless otherwise stated. Plots shown are typical representations of three experiments.

Strains and plasmids

A combination of standard molecular biology and molecular genetics techniques such as PCR-based tagging of endogenous genes and tetrad dissection was used to construct plasmids and strains (Supplemental Table S1). The plasmids for the CFP and YFP cassettes were obtained from the European *Saccharomyces cerevisiae* Archive for Functional Analysis. Further information regarding strain and plasmid constructions will be provided upon request.

Cell synchrony

For experiments requiring synchronized cultures, exponential-phase cells were diluted to 10^7 cells/ml in growth medium at 24°C. For G1 arrest, cells were treated with α-factor (US Biological, Swampscott, MA) at 0.4 µg/ml for 3 h. After the cells were arrested, they were washed by filtration and resuspended in media at the required conditions as described in the various sections. For a typical Noc arrest, cells were arrested with 7.5 µg/ml Noc (US Biological) for 2.5 h at 24°C, followed by the further addition of 7.5 µg/ml for another 2.5 h at 24 or 37°C, depending on the strain. The drug was washed off by centrifugation of the cells. Cells were then released and sampled at intervals as described in the relevant sections.

Fig. 7.2   Sub-sections in a Materials and Methods section.
(© Chin *et al.* (2012) *Molecular Biology of the Cell*, ASM.)

your lectures or when reading textbooks. For instance, common techniques encountered in Molecular and Cell Biology lectures are Western blot analysis, polymerase chain reactions and cell culture. Reading about the descriptions in the articles on such techniques might help you better appreciate what you have learnt in class that could be a little abstract.[18] The motivations behind the use of the various techniques will further be reinforced when you read the Results section subsequently. In the case when a technique is unfamiliar to you, do make it a point to clarify with your lecturers so that you would not have any lingering doubts in your mind as to what the authors were trying to achieve in executing various protocols.

What is Important in the Materials and Methods Section?

Many techniques are available to researchers to study biological processes, ranging from molecular biology tools to biochemical assays and genetic approaches. Other researchers might also make use of observational methods and computational modelling. As such, the researchers have no lack of tools with which to get at the answers of their research questions. The suitability of the techniques applied to the experiments is therefore a critical aspect to appraise when reading an article. It is the choice of tools and experiments that should be considered when you examine the article to decide if the appropriate strategy was taken and experiments were chosen to answer the research question. For example, you have to analyse whether appropriate techniques were used to

---

[18] I found that students learning about techniques such as these in lectures sometimes do not fully understand or appreciate the value or implications of using them. By highlighting common and relevant methods as part of my background preparation of students for the reading assignment, the students will get a more concrete view of how these techniques are useful for scientific investigations. Such preparatory work means that I explain a little more in detail the methods during my classes, with the inclusion of different examples taken from research articles to put the application of such techniques into a proper context.

investigate a particular biological process or if there were other techniques that were more suitable that could provide better answers to the research work.

You might have to read up on specific techniques if you have not already been taught in class. If you are not sure about a particular technique such as what the authors used it for, you could look up for more information or clarify with your lecturers. Otherwise, it is difficult for you to appreciate whether the technique used for the particular experiment is suitable and the significance of the data obtained. For instance, there might be more than one way to measure a particular phenomenon such as an enzyme activity. There could be different assays available to measure the particular enzyme activity. The choice would then depend on criteria such as the sensitivities of the assays. In the case of studying the sub-cellular localisation of a protein of interest, immunofluorescent staining of cells or live-cell imaging could be performed, although the choice of experiments would depend on whether the dynamics of protein localisation is required.[19] To ask such questions would depend on whether you are familiar with a range of techniques. Even if you are not, it would be good to simply bear in mind that there might be alternative methods of conducting the experiments described and to raise the question (e.g. among classmates) of whether there is another way the experiment could be performed.

By answering the questions provided in the Materials and Methods checklist (Table 7.1), you could again note down pertinent issues to help you focus on the objectives of the article. This list can be combined with the checklist on the Results section to provide you a better overall perspective of what the authors were trying to achieve in their research work.

---

[19] The notion is brought that alternative techniques for investigating a particular process is useful so as to introduce students to the idea that techniques are not prescribed for every experiment. Rather, they should appreciate that different techniques are available for answering a research question but that each might have specific advantages that could be more appropriate for investigating various aspects of a process. However, this depends on the level of maturity of the students and the time allocated for the class.

Table 7.1   Checklist for Examining the Materials and Methods Section

Questions	Remarks
Was the information provided in the Materials and Methods section detailed? Normally there should be details provided that would allow other researchers to replicate the experiments.	
Were there experiments that you were unclear about? What aspects did you not understand? Did you clarify with your lecturers?	
Are you able to determine the research strategy taken by the authors? This would require you to have an idea about the hypothesis as well. If at this point, it is not clear, that is not a problem. You might need to read the Results section and the accompanying data there to form an idea about the overarching research strategy. We will have more to say on this in the next chapter.	

8

The Results Section: What were the Important Observations Made?

The data described by the authors form the basis of the authors' stand concerning a specific issue they were researching on. That is to say, the support for the authors' ideas should come from the data resulting from their experimentation. Therefore, based on the data, the authors will be able to state their conclusions, which will be explored in the section on Discussion. In the main Results section, there would be in-depth elaboration on what data were obtained for each experiment conducted (Fig. 8.1).

In some cases, supporting evidence will be relegated to the section on Supplemental section if the data is not directly needed. For instance, the optimisation of certain experimental conditions for a new method or the initial calibration of a particular equipment could be published in such a section. In other instances, it could be data in the form of video files that are put online so readers could watch the videos to better appreciate dynamic events that are not easily reproduced in figures.

At the start of each sub-section of the Results, there could be some brief explanation as to why the authors chose to use certain techniques or experimental set-ups to answer a specific question. These are useful statements that help to lead into the sub-sections. The descriptive accounts of the results that follow are accompanied by the data, which are presented normally in the form of Figures or Tables, depending on the type of experiments that were executed.

51

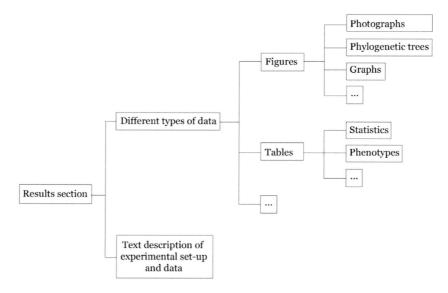

Fig. 8.1 Overview of the Results section.

The Figures and Tables will each have a legend that should describe the experiments in detail than what were in the general descriptions of protocols provided in the Materials and Methods section. For instance, experiment-specific conditions and set-ups that are essential for readers to follow the data will be stated in the legends. Other details such as deviations from the general description of protocols stated in the Materials and Methods section will also be included in the legend.

The Results section is divided into sub-sections with sub-headings (Fig. 8.2). Typically, the text in each sub-section describes the data presented in one figure. Each figure could have sub-parts (for example, A to E in Fig. 8.3), where the sub-parts are more often than not related to one another within the figure. For instance, the sub-parts of the figure could be from one complete experiment. Alternatively, the composite figure could comprise sub-parts that are data derived from different experiments but are placed together because they show a particular point.

You will note that at the end of each of the sub-sections in the Results, the authors will make a statement to sum up the key

RESULTS

Chs2–green fluorescent protein (GFP) fails to exit from the ER in the MEN *cdc14-3* mutant when mitotic Cdk1 is reduced

As part of the MEN, the primary functions of Cdc14 are promoting mitotic cyclin destruction by dephosphorylating Cdh1, and Cdk1 inhibition by dephosphorylating the Cdk1 inhibitor Sic1 and its transcription factor Swi5 (Visintin et al., 1998; Sullivan and Morgan, 2007). However, Cdc14 is believed to act more broadly by reversing phosphorylation on many other Cdk1 substrates (Stegmeier and Amon, 2004).

We examined the possible requirement for Cdc14 in Chs2 ER export independent of its role in promoting mitotic exit. To this end, we constructed a *cdc14-3 CHS2-GFP GAL-sic1-NTΔ* strain in which the overexpression of the truncated form of Sic1, sic1-NTΔ, allows the bypass of the mitotic exit defects in the *cdc14-3* cells (Mendenhall and Hodge, 1998; Noton and Diffley, 2000).

Of interest, in *cdc14-3 CHS2-GFP GAL-sic1-NTΔ* cells released

. . .

Chs2–yellow fluorescent protein (YFP) exit from the ER is defective in *cdc14* nuclear export mutant

We also examined Chs2 trafficking in a temperature-sensitive cdc14 nuclear export mutant (referred to as cdc14-NES) that fails to export Cdc14-NES to the cytoplasm normally. These cells however, are able to trigger the destruction of the mitotic cyclin Clb2 and enter the subsequent G1 (Bembenek et al., 2005).

In *CDC14 CHS2-YFP* cells released from Noc at 37°C, Chs2-YFP exited the ER with an increase in fluorescence signals at the neck and vacuole (Figure 1B, ii, 20 min onward). After Noc release, Chs2-YFP neck signals were seen in 27% of the cells at 30 min and 26% at 60 min (Figure 1B, iv). Moreover, after 100 min, 87% of the cells showing vacuolar signals could be observed with a corresponding loss of ER signals.

Fig. 8.2   Sub-sections in the Results.

(© Chin *et al.*, (2012) *Molecular Biology of the Cell*, ASM.)

finding of that particular set of data in the sub-section. Such summaries at the end of each sub-section help to link the data from one sub-part to another. Taken sequentially, these statements of key findings help the reader follow the progression of the data usually from some initial observations to more in-depth experimentation. The development of the authors' ideas through the sequence of experiments reflects the overall strategy of the authors' approach

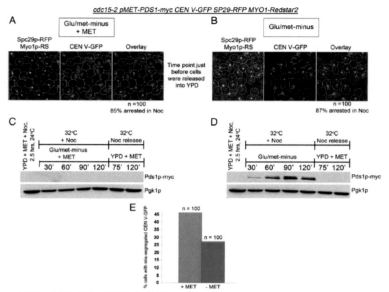

Figure 7. *cdc15-2 tetR-GFP CEN5::tetO2X112 SPC29-RFP MYO1-Redstar2 MET-PDS1myc* cells showed reduction in CEN V-GFP mis-segregation during Noc release when Pds1p was overproduced during Noc arrest. An overnight culture of the strain was arrested in Noc in YPD at 24°C for 2.5 h. The culture was then divided into two, with one-half of the culture resuspended in Glu/met− medium containing methionine (+Met) to repress the *MET* promoter driving *PDS1myc* expression and Noc. The second half was resuspended in methionine-minus medium (−Met) to de-repress the *MET* promoter in the presence of Noc. After 2 h at 32°C, both cultures were released into YPD. The cells after 2 h in Glu/met− with methionine (A) and Glu/met− (B) are shown together with the percentages of Noc-arrested cells. The percentages were determined from the number of cells from a total of 100 cells with both SPBs collapsed together as a single spot. Western blot analysis of the lysates taken at the indicated time points show the level of Pds1p-myc in the presence (C) or absence (D) of methionine. (E) Graph shows percentage of telophase cells with mis-segregated CEN V-GFP in +Met or −Met medium at 75 min after release from Noc.

Fig. 8.3   Example from a Results section showing figure legend and different sub-parts of a figure.

(© Chai *et al.* (2009) *Molecular Biology of the Cell*, ASM.)

(see Fig. 8.5). So if you record each of the main findings at the end of the sub-sections, it might be easier to follow the authors' logic in the research work described as you reach the end of the Results.

In some articles, you could find a figure (usually the last one in the article) that does not actually provide data but rather, that it presents a flowchart or a model in the form of a cartoon (Fig. 8.4). This is a common way that the authors use to try and sum up the findings in some sort of overview figure or schema. This can be quite useful for providing the readers a perspective of how the authors think the findings in the work link to other data in previously-published work.

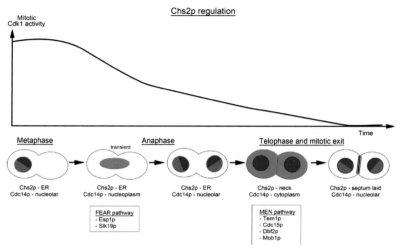

FIGURE 6: Schematic showing Cdc14 (green) and Chs2 (red) distribution during mitotic exit. See the text for details.

Fig. 8.4   Example of a figure in an article showing a schematic that puts the authors' findings in context with other previously published data. (© Chin *et al.* (2012) *Molecular Biology of the Cell*, ASM.)

In some of the articles you read, you might come across the phrase "unpublished data" or "data not shown" that the authors might have cited instead of referring to a particular figure or table when they describe a particular piece of data. In such cases, the data that are not shown in the article could refer to other findings in the authors' laboratory that are not directly relevant to the work described in the article. It is usual to see authors making references to the unpublished data perhaps once or twice at most in one article. The journals generally do not encourage multiple references to findings that are not published, as the article should comprise data that the readers can evaluate for themselves.

It could also be possible that the data that are not shown in the article have previously been included in the original manuscript that was evaluated by the peer-reviewers. But due to space constraints, the data was omitted from the final version of the article. Such data are again likely not to be directly relevant or critical for the work described in the article that you are reading.

Research Strategy Versus Individual Experiments

You should be mindful when reading the Results section (and the article as a whole) that there is an overall strategy taken by the researchers as compared to the individual techniques used for specific experiments (Fig. 8.5). The overall approach of the research project is critical as it sets the direction of the work such as the specific focus of the research project and the selected experiments to conduct.[20] However, the various techniques described in the Materials and Methods section are not in any way detailing the research strategy. Rather, it is a collection of the details relating to individual techniques used in the article. So getting at the research strategy does not mean merely reading the Materials and Methods section. Rather, it comes from understanding the hypothesis, appreciating the value of each technique and then examining the Results to form a perspective of the authors' overall approach to answering their biological question.

To distinguish between the research strategy versus the experiments, you could think of the strategy as the overall plan under which the authors have to design appropriate experiments and consequently, the techniques and protocols to adopt, to test their hypothesis. Collectively, the experiments form part of the authors' approach to answering their questions. As we shall discuss in the Results section, the organisation of the findings will provide a flow of the experiments that is consistent with the research strategy.

---

[20] For faculty members, it might be useful at this point to mention a little about what it means to the students the overall strategy versus various techniques used for experiments. Students might not be able to distinguish the two initially, depending on their exposure to research work. Giving them some ideas about the general strategy will help them learn how researchers approach a research question. Once the approach is selected, then specific techniques and experiments are selected. So researchers do not simply perform a series of unrelated experiments, but rather, the experiments are linked to one another to solve a biological puzzle when the findings are put together. Getting students to explore this aspect of research work would also highlight to them that the Materials and Methods section is a collection of techniques based on the researchers' strategy, though these are not arranged in the section in any order so as to present any sort of temporal sequence in which the experiments were conducted.

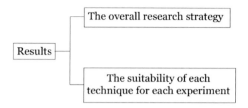

Fig. 8.5   The results can be judged on the general approach of the research work and individual experiments.

Basic questions you could start with when thinking about the general approach would be how the direction taken could address the research hypothesis. For instance, if the authors were keen to study the behaviour of a species of ant colonies, did they plan to carry our their observations at the natural habitat of the ants or did they isolate the ant colonies and study them in a laboratory setting and if so, what were their justifications for that. Once they had decided on this, what were the various experiments (i.e. techniques) they performed to observe the behaviour?

In another example, if a group of researchers were interested in understanding the effects of a particular drug on a specific cancer, did the researchers plan to conduct their study over a short- or long-term basis and what type of subjects would be included and how many? What specific experiments and techniques did they select to study drug effects on patients? Which statistical test did they use to determine if their data were significant?

In a more elaborate example provided below (Fig. 8.6), say a hypothetical research project described in an article aims to investigate the function of a novel protein X. That is the starting premise of the project. The overall strategy could be to focus on the role of protein X in regulating mitosis, using cells in culture. As part of the approach, different techniques would be applied to cells in culture to answer the question of what protein X might do during mitosis. For example, the use of small-interfering RNA to down-regulate the expression of protein X, the use of Western blot analysis to determine the expression profile of protein X over the cell division cycle, and immunofluorescence microscopy to determine the

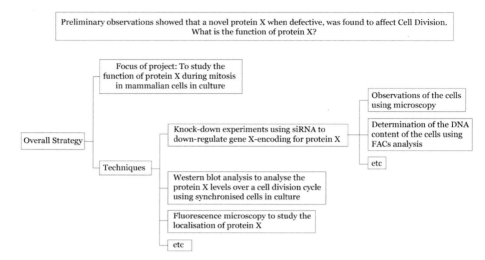

Fig. 8.6 An example of a hypothetical research project to highlight the point on the overall strategy of the research work as compared with individual experiments.

sub-cellular localisation of protein X are selected experiments that would collectively address the research question.

The general approach taken to study the research question will be discussed again later. This might then give you a more coherent picture of how the main bulk of the research article should come together when you examine the Results obtained from the researchers' approach and experiments.

Critique of the Results

It is pertinent to examine the technical quality of the experiments and results in order for you to assess the article appropriately. In essence, examine whether the choice of experiments or quality of data are good or not refer to how well they conform to the technical standards acceptable in the field. At the simplistic level, it could simply be to look at whether the data shown were unambiguous. For example, this could refer to whether bands on a Western blot

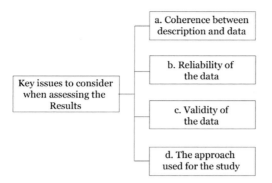

Fig. 8.7 Evaluating different aspects of the data described in the Results section.

or images provided for immunoflourescence staining of cells were clear. However, the issues in the Results section that could be examined in more depth are as shown in (Fig. 8.7).[21]

The technical considerations when doing research would include greater depth of analysis and would require you to think about the use of suitable controls for the experiments, data collection methods and strategies for analysis of results. The modes of evaluation and how far a conclusion one can draw from the data are also related to technical quality, though I will elaborate on these in the Discussion section.

a. *Coherence of data and description*

One other factor that could contribute to the quality of the data is whether there is good coherence between the data shown and what

---

[21] Breaking down the analysis of the data makes it less daunting for students, and at the same time, provides them some structure to build their analysis upon. Otherwise if the students were asked to evaluate the "quality" of the data without any guidance, the younger students would find it hard to even begin thinking about specific aspects of the data that they should assess.

the authors described. This normally requires you to scrutinise the data that have been presented in the figures or tables[22] and compare the actual data with that in the text in the Results section. This pertains more to whether the interpretation of the data by the authors is accurate.

At the initial level, you could study the data provided in the Figures or Tables and compare them with that in the text description provided in the Results section. There should be a correlation between the description of the data in the text and that provided in the figures. For example, if a Western blot is shown (Fig. 8.8), do you see the bands showing a protein of interest decreasing in intensities over a period of time if stated in the text that the protein was degraded as the experiment went on? Was comparison with a house-keeping protein such as actin whose levels are constant throughout the experiment made? Did they provide densitometric readings of the bands on the Western blot to show quantitative data on the difference? If the authors stated that there is a statistically significant change in the protein levels, did they provide densitometric readings of the bands on the Western blot and included details of statistical tests conducted to show significance? Is a statistical test needed?

In other instances, if the authors stated that a growth-defect phenotype was observed in a strain of mutant bacteria in a selective media as compared to wild-type strains, what type of data was presented (e.g. descriptive or quantitative)? Was the data provided in the figures consistent with the description?

---

[22] For students who are not experienced reading research articles, this is a useful tip for faculty members to share with students. For example, getting students to examine the data is a good way to train students to understand the experiments and the data. The students can also learn to be critical of data provided through such an exercise, as they have to pay attention to whether a piece of result is of sufficient quality and not accept what is stated at face-value.

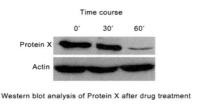

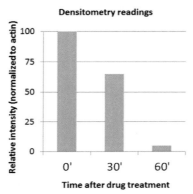

Fig. 8.8 Example of Western blot analysis of protein X with densitometry readings of the band intensities. Is there statistical test performed? Is it necessary?

b. *Reliability of the data*

i. *Are the data reproducible?*

Reliability means that the data should be reproducible within the study and by others when the same experimental set-up is used. Obviously it is not easy to judge how reproducible the data is by others but you could at least evaluate the data presented within the article for yourself based on simple criteria.

Take for example, experiments reported in an article comparing a normal enzyme with a mutant form of the enzyme using an enzymatic assay measuring the formation of a substrate (Fig. 8.9). You could look if triplicates of experiments were performed. If triplicates were performed, an indication as to the precision of the assay could be determined from the standard deviations among the

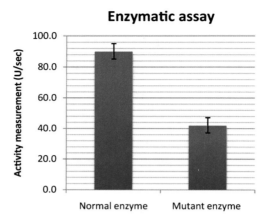

Fig. 8.9 An example of an enzymatic assay with small standard deviations.

triplicates. If the standard deviations for each of the enzyme and mutant assay were small, it would mean that the measurements were accurately performed and the data reproducible.

Statistical tests could be performed to determine if any differences observed were statistically significant. However, it should be stated that statistical significance does not equate with physiological significance. For instance, if there is a statistically significant difference in the enzymatic activities between a normal and a mutant enzyme as measured by an *in vitro* assay, does it mean that the mutant enzyme will cause any effects *in vivo*? There should be some other independent assay or test *in vivo* to determine if and how the mutant enzyme could adversely affect cells.

Conversely, you might sometimes find that control and test samples do not show statistically significant differences. This does not mean that there is no physiological difference unless this is confirmed by *in vivo* experiments. As such, you have to be mindful of how the authors interpret their data when looking at statistical analysis of quantitative data.

To further illustrate the point about reproducibility of data, take the instance of another pair of normal and mutant enzyme assay data: if the mean activity value for the normal enzyme was 90 U/sec and that for the mutant enzyme activity was 42 U/sec, and the standard deviations for both values were less than 10% of

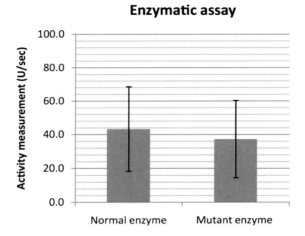

Fig. 8.10 An example of an enzymatic assay with low reproducibility.

the activities, then the authors could conclude that there was a difference between the normal and mutant enzyme. The authors could perform statistical analysis to determine if differences between the enzyme and the mutant form were of statistical significance. If the standard deviations were large for each of the normal (43 ± 25 U/sec) and mutant (37 ± 23 U/sec) activities, then the measurements were not reliable (Fig. 8.10), as the large deviations are an indication that the rates of enzymatic activity was relatively different among the three measurements taken. If the data were not reproducible in terms of the accuracy of the enzymatic assay each time it is run, then any conclusions made on the data would be weak. That is to say, with the large standard deviations (Fig. 8.10), it would not be clear which of the readings is the one that would properly reflect the true activity of the enzyme. These could have arisen from experimental errors, problems with the assay or random errors due to faults in the equipment. Therefore, you could not make a statement either confirming or refuting the idea that there is no difference between the normal and mutant enzyme.

For more descriptive or qualitative data such as the behaviour of the nematode worm, *Caenorhabditis elegans*, it might still be possible to enumerate such behavioural changes and provide quantitative information for ease of presentation and interpretation of

Table 8.1   Example of How Mutant Worm Phenotypes Could be Scored

Temperature shift from from 14°C to 26°C from 3 different experiments

Strain	No. of worms	% defects in forward movement	% defects in reverse movement
Wild-type N2 worms	88	1	0
Mutant 1	90	50	30
Mutant 2	85	2	76

the data. In cases such as trying to provide data concerning wild-type and mutant worm movements, it becomes necessary to note how the changes to worm movements are classified and scored. For instance, did the authors provide clear descriptors for the various phenotypes they used for scoring that would help them distinguish wild-type worms from mutant worms? For example, the movement of *C. elegans* worms could be scored based on their forward and reverse movements (Table 8.1). This would make it easier for the readers to understand what type of movement defects the authors were examining. Moreover, the data could be subject to more detailed analysis as there might be distinct genes controlling forward and reverse movements in the worms. Likewise, observations of other phenomenon would require researchers to state the criteria as to how the data were categorised unambiguously so the readers can interpret the data for themselves.

ii. *Is there consistency in the data?*

As part of the reliability of data, evaluations should also be made to determine whether different assays were performed to confirm a particular finding using a different technique. Taking the example of measuring the enzyme activity in a normal and mutant enzyme, were there data showing similar results with different assays? Why would this be vital to the overall quality of the data?

To illustrate this, let us assume that an enzyme that you are interested in is phospho-fructose kinase 1 (PFK1) which is needed for the conversion of fructose-6-phosphate and ATP to fructose-1,6-bisphosphate and ADP. This is the rate-limiting step for cells to commit to glycolysis. Assume for example, you wanted to measure the normal PFK1 activity and that of a mutant form of the $PFK1_{mut}$ to determine if the mutant form is more or less active as compared with the normal PFK1. You could determine the activities of both PFK1 using an *in vitro* enzymatic assay that measures the level of ADP formed by PFK1 as fructose-6-phosphate and ATP are converted to fructose-1,6-bisphosphate. This assay allows you to take lysates from cells harbouring the normal PFK1 or $PFK1_{mut}$ and use them for measuring the amount or rate of ADP formation as an indication of PFK1 or $PFK1_{mut}$ activities. Suppose that the assay indicated that $PFK1_{mut}$ has a lower rate of ADP formation. You could conclude that the mutant form as a lower activity than the normal PFK1. However, the assay was only one form of determination of the $PFK1_{mut}$ activity. Moreover, it is based on cell lysates, which might not reflect the $PFK1_{mut}$ activity in intact cells.

To test if the *in vitro* assay is consistent with the cellular activity of $PFK1_{mut}$, you could seek out other ways of measuring the PFK1 activity. As stated above, PFK1 is needed for driving glycolysis. An end product of glycolysis is lactate that is released into the extracellular environment. Lactate could be measured from the media taken off intact cells harbouring the normal PFK1 or $PFK1_{mut}$. If using such an assay revealed that cells with $PFK1_{mut}$ show lower levels of lactate than cells with normal PFK1, this would be consistent with the *in vitro* assay. In other words, the enzyme activities measured using an *in vitro* assay as well as using intact cells reflected a lowered activity in the mutant enzyme. This provides greater confidence to support the idea that the particular mutation in the enzyme can lead to defective enzymatic function. Figure 8.11 provides an overview of what has been discussed on the PFK1 measurement and consistency.

Another form of consistency is consistency of the findings to other reports in the field. For instance, there might be other

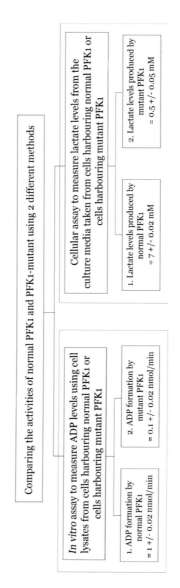

Fig. 8.11 Summary of example of PFK1 activity measurement.

published articles showing that a particular mutation in the PFK1 enzyme leads to reduced activity. As such, it would normally be expected that subsequent articles on this particular mutation of PFK1 reproduce this finding, unless there were errors in the original article.

In other instances when researchers talk about consistency, they are referring to how a particular observation can be confirmed or validated using different types of techniques. For instance, rather than relying solely on biochemical fractionation of cell lysates to study the sub-cellular localisation of a protein of interest, you could read about articles on immunofluorescence staining of cells using antibodies against the protein of interest, to support the data from the biochemical fractionation experiment. The strength of the authors' conclusions about the data becomes higher as there were different techniques used to examine their question.

c. *Validity of the data*

i. *Was there proper use of controls?*

In order for the authors to be able to make specific claims about their findings, it is essential that their data are valid. This can be seen from the point of view of whether suitable positive and negative controls[23] have been included in the experiments. Controls play an essential role in allowing experimenters to determine if a particular observation made was due, in fact, to random effects or errors, as the controls should provide baseline information on the assays or other forms of observations.

Positive controls are useful as they will indicate if the experiments were conducted properly. If the positive controls fail, then you would know that something has gone amiss, likely at the

---

[23] The importance of controls cannot be overstated. Some students doing research projects will invariably omit both positive and negative controls if they are not reminded to do so. So critiquing a research article can be a good alternative way to highlight to them the rationale behind using controls.

technical level. Without the positive control, you might not know that some error had occurred and you could be looking at a false negative result.

Negative controls are equally crucial in experiments, as they control for false positives. For instance, in a set of PCR reactions with controls and test samples, a negative control would be one reaction where no DNA template is added. This will show that the conditions and reagents used for setting up the PCR are clean. If your test samples show bands at the expected size but you also see a similarly-sized band in the negative reaction (such as arising from contamination and errors), you would consider carefully if the test samples were indeed showing a real positive reaction or you were looking at false positive result.

Take again the simple example of the PFK1 enzyme where we could measure ADP levels as an indication of enzyme activity as discussed above. To set up a more complete assay, controls such as the assay buffer and assay buffer with substrate controls (Fig. 8.12) are useful for showing the background readings measured by your particular instrument (e.g. spectrophotometer). Imagine if the controls gave very high background readings that you omitted to measure (such as due to a systematic error in the spectrophotometer), you might have results showing high readings for both the

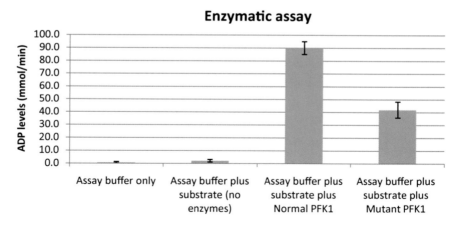

Fig. 8.12 PFK1 assay with controls.

normal PFK1 and the mutant PFK1. This might lead one to conclude that there is no difference between the normal PFK1 and the mutant PFK1, even though this could be a false negative result.

For such measurements of a specific end-product from enzyme-substrate reactions, it is also necessary to plot a standard curve of ADP concentrations. Such standard curves allow you to convert the measurements on a spectrophotometer to specific concentrations of ADP. The standard curve also helps provide an idea of the sensitivity of the assays. So if there is no difference detected between the normal and mutant enzymes, was the assay performed at the right level of sensitivity? So. Such questions can be raised as you examine the data from your article.

Take another instance of an article describing experiments relating to a drug trial where a new drug is tested on patients suffering from a specific disease. In a well-controlled study, there would at least be three groups of patients, one untreated group, one group given a placebo (a drug that looks like the new drug but without the active ingredient) and the last group given the new drug to be tested. It is well-known that a placebo effect, an effect of patients thinking that they have been given the active drug, can influence the patients' disease progression. Thus, even without being prescribed the active drug, patients on the placebo might also show signs of positive effects due to the placebo. Therefore, any positive effects due to the active drug will need to take into account the placebo effect. Any effects of a new drug over and above the baseline that is statistically significant would be of use when considering the use of the drug for therapy.

ii. *Are the findings validated using statistical analysis?*

Part of validity in quantitative data described in an article depends on the use of statistical analysis methods so as to determine if the findings were significant when comparing test samples with controls. As the use of statistical tools requires quite some explanation that is beyond the scope of this book, I only provide simple ideas for you to consider without going into the mathematical formulae.

For a more detailed description of statistical methods, you could refer to basic books on statistics.[24]

One consideration in the design of experiments that is aided by the use of statistical methods is the sample size that should be used in the experiments. The sample size could affect the validity of the findings, if the sample sizes used were too small, then any difference in the test and control cannot be accepted with confidence. Sample sizes are important for research work from laboratory-controlled experiments to clinical trials. It is generally easier to obtain sufficient number of samples in laboratory experiments than for clinical trials. For instance, it might sometimes be a challenge to recruit sufficient numbers of human subjects, or the numbers decline because the subjects drop out of the study mid-way.

If the sample size used for the experiments were sufficiently large, then you would have to consider if the quantitative differences observed in the data between the test and control samples were statistically significance. Usually, there would at least be some form of statistical tools used determine the probability (p value) that the differences were due to random effects. Normally values such as $0.01 < p < 0.05$ are considered statistically significant, meaning to say that the probability is very low that the differences observed between the test and control samples were due purely to coincidence. A range of methods is available for calculating the p values, depending on the nature of the data. One question you might want to bear in mind is whether a statistically significant difference is always equivalent to a biological or physiological significance.

Other than testing for significant differences, there are also statistical tests for determining correlation between variables. Take for instance, an experiment set up to study the effects of age on the rate of photosynthesis in a plant. So there are two variables, the age of the

---

[24] Some universities would provide basic statistics courses for Life Sciences students. For those that do not, it might be useful to go through some general ideas without having to delve too deeply into this. I would normally explain what is necessary for students to understand the analysis done in the article of interest. Otherwise, some students might get too confused and be distracted from the rest of the article.

plant and the rate of photosynthesis that the experiments should measure. If the observations were that the older the plant, the slower the rate of photosynthesis, statistical tests could be performed that would help to determine if this correlation was strong or weak.

While there are tools to provide some sense of whether the data is of statistical worth that would give more confidence to the results, you should distinguish this from biological significance. As mentioned above, showing statistical significance between test and control samples does not necessarily mean that there is a biological consequence due to the test conditions.

iii. *Are experiments selected to answer the questions that the authors set out to investigate?*

Last be certainly not least, a very crucial question relating to the validity of the data is whether the experiments the authors designed actually measure what they set out to determine. This would mean looking at the research questions and data from the experiments and judging if the data obtained from the experiments answered the questions directly.

Take for example the case of temperature effects on *coral* asexual reproduction as mentioned in Chapter 6. If the measurement of asexual reproduction is based on the number of new offspring found within a distinct radius in a given time for various temperatures, would you consider that the enumeration of new offspring a good method to study whether temperature affected reproduction *per se?*[25] The choice of experiments or assays to conduct therefore

---

[25] This aspect of critically examining the data with respect to the actual research question is one of the more important objectives of getting students to learn to critically evaluate research articles. Often times, students gloss over the analysis of the rationale behind executing the experiments and accept what is stated in the textbooks or articles without pausing to consider carefully the validity of experiments. They might not be familiar with questioning the "authority" of the scientific article, so letting them practise querying the basis of the experiments is a useful exercise.

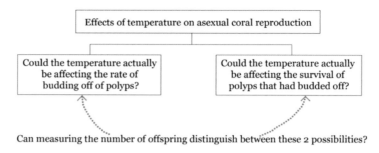

Can measuring the number of offspring distinguish between these 2 possibilities?

Fig. 8.13   Is the experiment measuring the correct process?

needs to be examined closely as the data collected would shape the conclusions made by the authors. This will be further elaborated on in the Discussion section.

At the surface, it might seem so, as the offspring are the products of a reproductive process. However, if you consider that the number of offspring might reflect different issues, then the experiment to merely measure the number of offspring as an indication of asexual reproduction might be inaccurate (Fig. 8.13). To study how the temperature affected asexual reproduction, one could measure the rate of budding of polyps from the main coral head as a direct indication of reproduction. One could also track the survival of the polyps that had budded off, as the temperature might affect the viability of the newly-budded of polyps that could influence the number of offspring surrounding the parent coral that you count. So would merely counting the number of offspring give you a direct answer as to how temperature might impact reproduction?

d. *The approach used for the study*

i. *Strengths and weaknesses of the approach taken*

As mentioned in the Results section, the overall strategy (Fig. 8.5) that the authors used to approach their question could also play a part in the quality of the data as it relates to the choice of the experiments included in the article. In turn, the choice of the article will result in a distinct collection of data upon which the authors will make their conclusions.

Typically there are different pathways to answering biological questions.[26] This can sometimes be seen quite obviously in articles published together by separate groups in the same issue of a journal where the groups made use of different approaches but reached similar conclusions.

For example, in the characterisation of the mitosis-promoting factor that is needed for triggering mitotic events, two different groups published back-to-back articles in 1988 showing the active component of the complex using different approaches: Dunphy, W. G., Brizuela, L., Beach, D., and Newport, J. on "*The Xenopus cdc2 protein is a component of MPF, a cytoplasmic regulator of mitosis,*" and Gautier, J., Norbury, C., Lohka, M., Nurse, P., and Maller, J. on "*Purified maturation-promoting factor contains the product of a Xenopus homolog of the fission yeast cell cycle control gene cdc2+*" (Dunphy, Brizuela, Beach, and Newport, 1988; Gautier, Norbury, Lohka, Nurse, and Maller, 1988).

When you examine the strategy adopted in an article, you could consider whether the strategy taken was too narrow or too board to address the authors' questions. Did the choice of the experiments restrict the scope of the study such that only a small part of the research question was answered. In such articles, perhaps you might notice that there is more depth in the type of data that the authors could provide such as detailed mechanism of a specific enzyme.

In other articles where the approach might be rather broad such as investigating the different homologues of a particular enzyme in a particular genus of bacteria, then the study is likely to be more of a survey than providing mechanism of action of the enzyme in question. By questioning the approach employed in the

---

[26] This is a pertinent point that could be brought to students' attention as they learn about the process of scientific enquiry. It would be important to give them the idea that scientific research that follows a strict set of rules and methodologies, is all about using creativity and innovative ways to solve a biological puzzle. As such, different strategies have been successful at coming to similar conclusions.

study, you could better appreciate the nature of the data obtained and how they could address the authors' research question. Also, this could help you put the authors' conclusion into perspectives (see later in Discussion).

Whether the strategy taken is broad or narrow does not necessarily make the article a good or a bad one. The issue is whether the main research question set up by the authors was answered by the experiments. This includes the validity of the data. Moreover, the quality of the data in terms of reliability plays a role in the assessment of the strengths or weaknesses of the article.

ii. *Suitability of the experiments*

There are different approaches a research question can be answered as seen in the example given above on solving the active components of the mitosis-promoting factor by two different research groups. Likewise, at the experimental level, there are normally different techniques[27] available to researchers when they want to examine a particular biological process. The purpose of assessing the experiments is to consider if there were alternatives that could be more appropriate than others in arriving at data that could be relevant to the research question. Take the instance of studying the possible changes in the localisation of a protein of interest during cell division. Would experiments using logarithmically-growing cell culture to study the protein localisation be more or less appropriate than using synchronised cells in culture? Would the study of genome-wide expression patterns of metabolic genes during drug treatment of a cohort of diabetic patients be more relevant than the study of a few key genes?

A summary of what you should look out for in the Results section is shown in Fig. 8.14. This might seem overwhelming if you are reading an article for the first time, but if you break down your

---

[27] This requires students to know more than one technique to study a particular process. As mentioned in footnote 19, the rationale for highlighting that different techniques are available for studying a particular process is important. This should be emphasised so that students understand the reasons for choosing a particular technique over another and not learn by memorising a list of techniques.

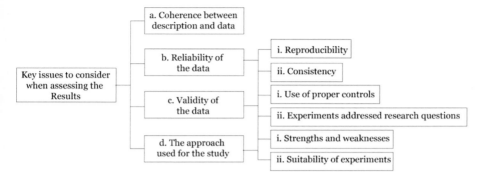

Fig. 8.14 An overview of the key factors to assess in the Results section.

Table 8.2 Checklist to Help Students Go Through the Results Section

Questions	Remarks
What were the key findings reported by the authors?	For each figure in the article you are reading:
• Figure 1	How would you judge the reliability of the data?
• Figure 2	
• ...	What do you think about the validity of the results provided?
Was the approach used for the study good?	Key strengths:
	Key weaknesses:

analysis of each sub-section or Figure at a time it would not be that difficult. It might also be productive to do this with a class-mate or two so that you could help one another if there are doubts. But over time as you become more experienced reading scientific literature, this will become second-nature to you as the key criteria for different articles are largely based on these main guidelines.

To further assist you in the evaluation of a research article, you can make use of the checklist shown in Table 8.2 to consolidate the essential details of the data reported in the article. Filling in the table in a progressive and systematic manner, you will develop a better understanding of the article as you have to think through what each part of the Results or data mean. So by the time you get to the end of the Results section, you would probably have some opinion of your own about the article.

9

The Discussion Section: What were the Main Conclusion(s) Made by the Authors Arising from the Data?

The Discussion section is where the authors elaborate on their analysis of the data described in the Results section. This is different from the Results section in that the description in the Results only deals with the findings and the conditions for specific experiments. In the Discussion, in addition to data analysis, the authors could further attempt to put the data into the context of the research question (Fig. 9.1). The authors also have to make justifications as to whether the data support their initial hypothesis and how their findings help to expand the existing knowledge. Normally, the authors would make comparisons of their data with what is currently known in the field and then make statements as to whether their results are consistent with what is accepted. At times, they would even draw up a model (Fig. 8.4) in an attempt to provide a synthesis of their ideas in light of the existing knowledge. If that is the case, the findings might serve to extend what is known, and as such, help to deepen the knowledge in the field.[28]

---

[28]The students should learn that the Discussion section is not simply a summary section for the Results. Oftentimes, junior students writing up a thesis for an undergraduate research project make use of the Discussion section as a place to repeat their description of their data. This makes me wonder if they did not really read the research articles carefully to note that the Discussion section actually is

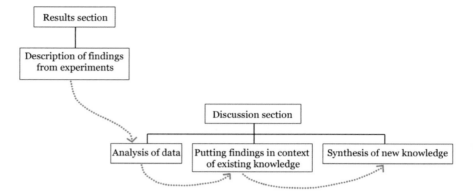

Fig. 9.1   Discussion section.

In the case that the data are not consistent with the current accepted understanding in the field, the authors have to make a case as to why their data are contradictory. They would have to exclude the possibility of errors due to experimentation, by pointing out the reliability and validity of their experimental set-up, and bring up other reasons that could explain the discrepancies between their data and the current knowledge. Improved technologies leading to better acquisition and/or analysis of data could result in a change in the understanding of a particular biological phenomenon. The question is whether the other researchers in the field are convinced of the new ideas presented in the article.

The authors then make their conclusions in the final paragraphs of the Discussion section. Usually, the authors will also discuss possible weaknesses of their study and offer solutions for subsequent studies. In addition, the authors might end off with further suggestions as to the possible work that could be performed in the near future to extend the current work.

---

where the authors try to explain their findings in relation to what is already known and how their work extends the current knowledge. It would therefore be worthwhile to highlight this point to them.

Critique of the Conclusions

a. *What were the significant points of the findings in the article?*

The key findings in the article would be explicitly stated by the authors in the Discussion section. More essentially, the authors have to discuss the data with respect to how their data fit into what is currently known in the field. At the end of your reading, you could ask yourself what you think were the key pieces of data that were reported in the article. This information can be obtained from your own analysis of the data presented in the various Results sub-sections. As mentioned before, you could note down what the take home message is for each figure so as to break it down into smaller chunks of information as shown in Table 9.1. Once you have done that for individual figures, you can combine them together to get a better overall idea of what the significant findings of the article were.

You could compare what the authors have stated with your own analysis of the data and ask whether the findings from the paper have indeed contributed new and interesting ideas to what is currently known in the field. For example, the authors could have data indicating that the cell division control in a mammalian cell in culture is similar to that in a simple eukaryotic budding yeast cell. Alternatively, the authors might have data showing new mechanistic insights into the transcriptional control of a known gene through the characterisation of a new protein and its contribution to the transcription pathway. Such articles providing data that shed new light on particular fields of research contribute towards extending the knowledge in the field.

At times, the data published in articles could contradict[29] the existing knowledge in the field. The findings from these studies are

---

[29] As mentioned above, contradictory ideas or findings are interesting to discuss with students. It is useful to highlight the fact that scientific knowledge is tentative and subject to change. This is part of teaching students about the nature of science

Table 9.1   Checklist to Help Students Summarise the Discussion Section

Questions	Remarks
What were the main conclusions made by the authors? You could break down the conclusions into the key idea represented by the data from each figure: • Figure 1 • Figure 2 • Figure 3 • …	How does the main conclusion from each figure lead to the next figure? Do you agree with the authors' conclusions?
Did the data answer the overall research questions posed by the authors?	
What did you think were particularly convincing?	
Were were the gaps in the work?	

no less significant, as such alternative perspectives might lead researchers to re-examine the knowledge that they hold and perhaps re-look at their understanding of their subject. Contradictory data sometimes come about because of the advent of new technologies that allow researchers to re-investigate old issues or study new questions that then provide new information leading to a different understanding of the topic.

and should help them appreciate that learning to consider different points of views on a theory is an important aspect of a university education. Such topics can also persuade students to stop thinking about biological knowledge as a fixed body of information that should be learnt through memorisation.

b. *Did the authors make the right conclusions based on their interpretation of their data?*

Judging from the quality, approach used and the extent of the results, was the conclusion stated by the authors supported? To get at how the authors make the conclusions they did, one way is to find out how the data from each sub-section in the Results section were linked to the next sub-section. The flow of the data should follow a logical progression, usually from the broad to the narrow if the authors were working on finding out details of a mechanism such as from the interaction between two proteins to their interaction domains to specific amino-acid residues that are involved. Alternatively, the study may involve a biochemical pathway from a particular species of fungus and extending to its related members. To establish the links, you could note down what each key conclusion from each sub-section of the Results section is and then ask yourself how that leads to the next experiments (Table 9.1).

Next, you have to consider whether you agree with the authors' conclusions based on your own examination of their data in the figures and tables. In some instances, the authors might over-state their conclusions by extrapolating a little too much from their data. So the crux of the issue is whether the statements made by the authors were directly supported by their data.

Using the hypothetical example shown in Fig. 8.6 on the study of a novel protein X and its role in mammalian cell division, supposing that the data collected from the experiments are as shown in Fig. 9.2. From the experiments using siRNA (steps 1a and 1b in Fig. 9.2), the authors observed that the absence of protein X led to cells stopping cell division with 4N-DNA content. This meant that cells failed to divide after they had replicated their DNA. Based on these findings, the authors concluded that protein X has a role to play in the progression of the mammalian cells in the post-DNA replication phase. This statement would be supported by their data.

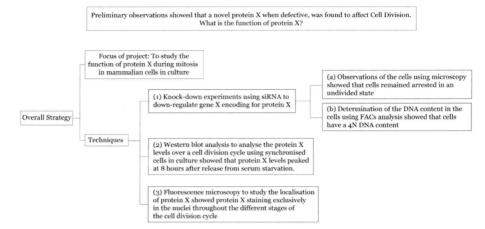

Fig. 9.2   Results from hypothetic research project.

From the other pieces of data, if the authors concluded that protein X played a role in regulating cell division through gene transcription as immunofluorescence-staining of protein X indicated that protein X is a nuclear protein (step 3 in Fig. 9.2), then the statement would not be correct. This is because a nuclear protein is not necessarily a transcription factor or a protein that has a function related to transcription. So making inference from the observation that protein X in the nucleus to its involvement in transcription is over-stating the fact. This rather obvious example used here is to illustrate the point about whether conclusions are directly substantiated by the data. But the take home message is that when you are reading the conclusions, you need to re-look the data and be clear that the interpretation of the data is precise and not extrapolated from weak links to the data. This would help you decide if the data were consistent with the conclusions stated by the authors.

c. *Did the data answer the questions/hypothesis raised by the authors?*

Related to the section above is the question of whether the authors managed to gather sufficient information to address the hypothesis they raised at the beginning of the article. It might again be worth-while to revisit your notes on what the authors' research question was that had been stated in the Introduction section and compare that with the evidence provided in the Results section and the con-clusions offered in the Discussion section. In most cases, the data could answer the hypothesis posed at the beginning of the article, although, usually the data might not be complete as there is not likely to be a perfect approach or a technique available to solve each question.

You could also see if the authors offered any evaluations of their findings. Did they provide a fair evaluation of the strengths and weaknesses of their data? In the case that their data deviated from their original hypothesis or predictions, did they provide alterna-tive explanations for the findings? Was the explanation provided logical?

There could also be new questions arising from the study that you might be able to think of. This is not too surprising, as new issues often arise from a research project. Also, normally a single article would not be able to solve major questions in the research topic. Rather, it requires the scientific community in the research fields many years of accumulating sufficient data to provide a big leap in the advancement of a biological process. Nonetheless, the efforts of the researchers as seen in each paper would require at least a few years' work on the part of a postgraduate student and/ or postdoctoral fellow.

So it could be possible to think about what gaps[30] there are in the current study and how you could design experiments to further the study. Of course, whether this is a requirement in your assign-ment depends on your lecturer. But as a matter of practice, you

---

[30] I sometimes incorporate such activities in the assignment to try and encourage students to think beyond what they have read. This is especially useful for a small part of the assignment that I award a bonus point to cater to more advanced stu-dents in the class who need some stimulating or challenging activity. As the amount of bonus marks awarded is small, the low risk nature of this exercise would even persuade students who are not that confident to provide some suggestions.

could discuss with your classmates ideas that you have come up with. You could find out more about whether the authors extended their findings subsequently by searching in Pubmed, other articles published by the authors subsequent to the article that you were reading.

The checklist in Table 9.1 will help you summarise the main ideas in the Discussion section. Upon reaching the end of the Discussion section, you could re-visit the various sections to check if you need to clarify any outstanding doubts about the work. One benefit of keeping checklists (such as those I have suggested) on specific queries you might have is that you can re-examine those questions you have initially and find out if you have gained a better understanding after going through the authors' explanations in the Discussion.

10

What are Your Views on the Article?

At the end of your reading, you would have examined the details of the various aspects of the article. So are you able now to state what you think the article[31]? For instance, you might or might not agree with the strategy of the research, the interpretation of the results and the conclusions stated by the authors. The prime objective of reading an article critically is to learn to form your opinion of the authors' data presented in the article by going through each section and systematically evaluating what has been presented. This way, you can back up your views with examples of the strengths and weaknesses in the article and not merely state your support for or rejection of the article based on guesswork or "gut-feeling." That is to say, you should move away from making a stand without a basis.

In essence, scientific assessment of a research article very much relies on justified statements supported by evidence. The support of your statements comes from the factors you have made in the various checklists provided in the book. Of course the checklists can be modified to suit your own preferences and requirements if

---

[31] Getting students to state their own views can help encourage independent thinking and reinforce the idea that they should not simply rely on the printed word as the correct source of information. If the students have gone through the article systematically, it is possible for them to develop some thoughts they might have on the article, even if they do eventually agree with what the authors have stated. So organising a discussion session with the students or an online forum among themselves could be a possible activity to carry this out. A summing up session such as at a class tutorial can then be set up to conclude the discussion of the article.

needed. Nonetheless, going through the exercise of actively evaluating each section of the article will equip you with the basis for forming your opinion of the work.

For instance, if you disagreed with the authors' conclusions, could you provide a reason why? Was it a problem with the reliability of the experiment? Could there be an issue with the validity of the experiment? Was it a problem of interpretation of the data by the authors? Whichever it is, you could substantiate your opinion with concrete examples of which aspect of the article you deemed were weaknesses. Conversely, if you agreed with the authors' conclusions, do you have sufficient grounds to support your views? Thus, when you reach the end of the article after carefully and critically assessing the information provided, you would likely have an overview of the article that comprises not only of the main points made by the authors, but also your own perspective of the work (Fig. 10.1). You could then discuss whether you think the findings reported are of sufficient quality and rigor and if you agree with the conclusions made in the articles based on your examination of their data.

You could also form some opinion as to the other aspects of the article such as whether the authors were able to present their ideas coherently and logically. Were they able to explain their hypothesis and experimental designs in a manner that made it easy for you to read the article? Also, was the presentation of the data straightforward with clear legends or was it complicated and needed repeated reading of the text before you could comprehend the data? These factors though not directly related to the experimental work, could collectively influence how readers respond to an article, as a badly presented piece of work will make it difficult for them to read it objectively (you should bear this in mind when you have to write

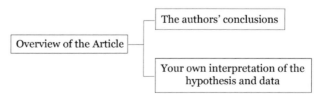

Fig. 10.1 Providing your own perspective on the article after reading it.

an essay such as writing a critique of a research article — see next section). Table 10.1 shows an example of a checklist that can help you consolidate your thoughts on the article.

As you progress through your undergraduate work, being able to deconstruct a research article in a methodical manner is beneficial for analytic work related to research projects and thesis writing. Moreover, such skills are transferable to other scholastic activities that need critical reading and analysis. This will obviously take some time to develop, so it requires patience and persistence when reading through your first research article. However, once you have acquired the ability to evaluate information in a systematic way, it will remain with you and serve you well even after you leave university.

Table 10.1   What are Your Thoughts on the Article

Questions	Remarks
What were the main conclusions made by the authors?	Do you agree with their conclusions?
Conclusion 1	Agree/Disagree
	Reasons:
Conclusion 2	Agree/Disagree
	Reasons:
Overall presentation of the work	Writing
	Data presentation

11

Writing Activities Related
to Critiquing an Article

Depending on why you have to read a research article, you might have other activities that follow your reading activity. For instance, based on reading a selected research article, you might have to write[32] a commentary article for a term paper, an argumentative essay for an assessment or a literature survey for a project work. These various writing assignments will need you to apply critical analysis of the research article as well as write down your thoughts in the form of an essay. The styles of writing might be different, as different tasks will have specific emphasis and requirements.

For those who need some guidelines on how to write an essay to critique or review an article, the following points might be worthwhile for you to consider as you plan your writing. Planning[33] is needed

---

[32] Writing-to-learn is useful to reinforce students' critical examination of the research articles published in scientific journals. Much like getting students to discuss their views about the articles (see footnote 31), by writing about the article they are reading, students get a chance to examine deeper, the evidence in the article. They will have to consider if the data were convincing to them and their peers in their own writing and this will engage high-order thinking skills as shown in previous studies cited in the Preamble.

[33] Faculty members could encourage students to use a Mind-map, a scrap of paper or any other media they prefer to plan out their essays. It is very useful for students to have an outline to organise their thoughts before typing out their essays. That way, they are encouraged to read through the article with a critical eye as they have to explicitly plan out what they want to include in their essays.

before writing as you need to know what the key points are you would like to discuss and how your ideas should be organised in the essay.

For instance, you could make use of what we have discussed thus far to plan your essay (Table 11.1) if you have to write a critique of a selected article for your assignment. The objectives are to systematically evaluate the research article you have to read by going through the steps I have described above and then write down the salient points in a logical manner.

Essentially, to write your critique of an article in the form an essay, you should start with an Introduction section to provide an overview of the topic in the article. In your Introduction, you could learn to make use of citations to support your statements if you have read a little beyond the article you have to critique — whether this is required depends on the nature of your assignment.

Table 11.1   Tips on How to Write an Essay Critiquing a Research Article

Key points in the essay	Remarks
Write an Introduction to give a brief overview of what is known about the subject matter.	Refer to the section on Introduction. More advanced students might refer to review articles to provide a broader scope.
State the key issue(s) you want to focus on.	Refer to the section on Introduction.
Explain what hypothesis the authors of the article proposed to test. Describe the approach taken by the authors.	Refer to the section on Introduction.
State the data obtained by the authors.	Refer to the section on Results.
State the conclusions drawn by the authors based on their interpretation of their data.	Refer to the section on Discussion.

(Continued)

Table 11.1   (*Continued*)

Key points in the essay	Remarks
Evaluate the article in terms of what we have discussed in the Chapters above.	
• The hypothesis — Comment on the relevance of the hypothesis in the context of the knowledge at that time.	Refer to the section on Introduction.
• Technical quality — refer to the section on Results:	Refer to the section on Results.
○ Was the approach taken good?	
○ Were the data reproducible?	
○ Any issues with validity of data?	
○ Consistency within the article — for example, how large were the error bars in a graph? Any statistical analysis performed? Significance values?	
• Consistency of the data in this article compared with what is known from other reports — this is more likely to be important when you do your research project and are writing up your literature review, when you will be reading more than one article.	Refer to the section on Discussion.
• Did the authors make the right conclusions based on their interpretation of their own data?	
• Did the data answer (or not) the questions /hypothesis raised by the authors?	

(*Continued*)

Table 11.1   (*Continued*)

Key points in the essay	Remarks
Discuss briefly whether you agree with their conclusions based on your examination of their data.	
Conclude with a statement on what you think overall of the article.	

Then there could be a main body where you present your critique on the various aspects of the article such as the hypothesis, Materials and Methods and Results. For the critique on the Results, as you analyse each figure, you could cite the relevant figure number to make clear which aspect of the data you are evaluating so the readers can follow.

Towards the end of the essay, you could discuss whether you agree with the authors' conclusions, using the various analyses you have made to substantiate your views. You can then elaborate on whether the article has or has not answered the authors' hypothesis and other further areas in the research that could be improved and how. Finally, you could make a general statement on the impact of the findings in the article in the field.

Note that not all the points on writing a critique are applicable to other forms of essays as it depends on the objectives of your assignment. Also, the layout of the essay could be different from what I have listed, as other factors such as your own preference and style of writing and of course, the guidelines for your assignment at hand. However, the suggested points can be easily adapted for other writing assignments such as a literature survey or a short commentary.

Nonetheless, you could probably see that the skills needed for the reading and critiquing a research article is closely linked to other learning activities such as writing[34] an essay on a research

---

[34]With regard to essay writing, students who have not learnt English as a first language might have some difficulties even with basic grammar, sentence construction

article. You could make use of the principles involved in critiquing an article for other activities such as evaluating the aims, evidence and conclusions of other information that might be presented to you in newspaper articles, debates among classmates as well as data on other issues relating to social matters that you might come across in policy reports. Hence equipping yourself with the ability to evaluate a scientific or academic article could be valuable for other aspects of your studies or professional life that need a critical eye, as the skills are broadly applicable.

---

or even understanding the article. For some universities, there could be specialised English language courses that could provide some guidance and support. These students could make use of such courses to help themselves at least improve upon their general writing skills. For students without such a support, they might need to approach the faculty members teaching the Life Sciences module for a bit more help. When I grade essays from students who are not used to writing in English, I try to focus on the content more than the students' language abilities, though it is sometimes challenging as having a good grasp of the English language is an advantage when the assignment is essay-based. Instead of writing a full essay, I sometimes design assignments based on reading research articles that require only short answers.

12

Final Words

The reading of scientific articles can be a challenging activity especially if you are doing it for the first time. From my experience using reading and writing assignments as part of learning and (fairly low-stakes) assessments, students who responded to my end of semester surveys reflected that they found the reading not easy at the beginning. However, they noted the tips I provided for reading articles helped them navigate around the research article. Moreover, once they get past the initial difficulties such as the unfamiliarity in handling scientific articles, they actually thought the exercise beneficial in several ways.

Among the more striking rewards of learning to read scientific articles was overcoming their initial apprehension when encountering a research paper. In addition, the confidence they gained after going through the first article was useful when they had to deal with subsequent reading assignments involving yet other scientific articles. In fact, those who responded stated that they were surprised they could actually understand the articles I set for their reading tasks. Other benefits students indicated they gained through learning to read a research article was that they understood a little better the process of scientific discovery.[35] There are also the more practical

---

[35] It should be noted that depending on the time allocated for such reading assignments, it might not be possible to discuss with students all that is important about the scientific process. It also depends to a certain extent on the maturity of the students. For instance, I find that I did not have time to elaborate on issues related

skills of using databases to search for articles and handling literature survey for their project work.

A relevant point for students is that the reading of scientific articles might become more and more part of the science curriculum because of the usefulness of incorporating them as teaching materials. So for students aiming to continue on to graduate studies, learning to read scientific articles early on in your studies will no doubt allow you more time to find out more about the scientific literature that primarily deals with empirical research data. The goal is to make use of the content knowledge you have learnt in class in your reading of the literature. In addition, you should learn to acquire skills that might not be taught during lectures, including critical analysis of the authors' claims whether in their hypothesis or discussion of data. Some of these skills will take time and effort to develop, so just have a little patience when learning about the scientific enterprise through reading scientific articles.

For students who might not continue on to graduate school, it is still a worthwhile effort picking up skills to critique the data in a scientific report. These skills are transferable to other aspects of your lives such as in your profession when you have to evaluate information provided to you by clients and colleagues. The ability to critically analyse figures, statistics and other forms of materials such as in a news report on the effects of a new drug or information on a disease using a systematic and evidence-based approach could help you in your personal lives as well.[36]

---

to the nature of science but could only touch on this topic at a surface level with my second year large-class (280 students) module. Nonetheless, the use of scientific literature is actually a good point to start some discourse on the nature of science that is useful for improving students' view of scientific research and perhaps even the nature of scientific knowledge.

[36] Students should not regard learning how to read scientific articles as a skill only useful for those moving on to science-related careers. It should be impressed upon them that the skill to evaluate data or information in a research article is widely applicable and so they should try to make use of the opportunity to learn what they can and practise on other forms of reading materials.

In conclusion, scientific experimentation is the path to building scientific knowledge. The enterprise of the scientific research therefore revolves around the publication of experimental data for the Life Sciences research community to communicate, critique and hence extend what it is that is known about the disparate biological phenomenon. As a science student learning about the workings of the biological world, reading about the research process and how it is conducted through reading scientific articles will help in part in contributing to your understanding how the body of scientific knowledge comes about. So picking up the skills to read the literature that is closely linked to scientific knowledge can only help you better appreciate how we know what we know about life sciences today.

Bibliography

Asai, D. J. (2011). Measuring student development. *Science* (New York, N.Y.), **332**(6032): 895. doi:10.1126/science.1207680.

Chai, C. C., Teh, E. M., and Yeong, F. M. (2010). Unrestrained spindle elongation during recovery from spindle checkpoint activation in cdc15-2 cells results in mis-segregation of chromosomes. **21**, 2384–2398. doi:10.1091/mbc.E09.

Chin, C. F., Bennett, A. M., Ma, W. K., Hall, M. C., and Yeong, F. M. (2012). Dependence of Chs2 ER export on dephosphorylation by cytoplasmic Cdc14 ensures that septum formation follows mitosis. *Molec Biol Cell* **23**(1): 45–58. doi:10.1091/mbc.E11-05-0434.

Deboer, G. E. (2000). Scientific Literacy: Another look at its historical and contemporary meanings and its relationship to science education reform. *J. Res Sci Teach* **37**(6): 582–601.

Dunphy, W. G., Brizuela, L., Beach, D., and Newport, J. (1988). The Xenopus cdc2 protein is a component of MPF, a cytoplasmic regulator of mitosis. *Cell* **54**(3): 423–431. Retrieved from http://www.ncbi.nlm.nih.gov/pubmed/3293802.

Gautier, J., Norbury, C., Lohka, M., Nurse, P., and Maller, J. (1988). Purified maturation-promoting factor contains the product of a Xenopus homolog of the fission yeast cell cycle control gene cdc2+. *Cell* **54**(3): 433–439. Retrieved from http://www.ncbi.nlm.nih.gov/pubmed/3293803.

Green, W., Hammer, S., and Star, C. (2009). Facing up to the challenge: Why is it so hard to develop graduate attributes? *Hig Edu Res Dev* **28**(1): 17–29. doi:10.1080/07294360802444339.

Hagting, A., Karlsson, C., Clute, P., Jackman, M., and Pines, J. (1998). MPF localization is controlled by nuclear export. *EMBO J* **17**(14): 4127–4138. doi:10.1093/emboj/17.14.4127.

Hoskins, S. G., Lopatto, D., and Stevens, L. M. (2011). The C.R.E.A.T.E. approach to primary literature shifts undergraduates' self-assessed ability to read and analyze journal articles, attitudes about science, and epistemological beliefs. *CBE Life Sci Edu* **10**(4): 368–378. doi:10.1187/cbe.11-03-0027.

Hoskins, S. G., Stevens, L. M., and Nehm, R. H. (2007). Selective use of the primary literature transforms the classroom into a virtual laboratory. *Genetics* **176**(3): 1381–1389. doi:10.1534/genetics.107.071183.

Hurd, P. D. (1998). Scientific Literacy: New Minds for a Changing World. *Sci Ed* **82**(3), 407–416.

Kozeracki, C. A., Carey, M. F., Colicelli, J., and Levis-Fitzgerald, M. (2006). An Intensive Primary-Literature — based Teaching Program Directly Benefits Undergraduate Science Majors and Facilitates Their Transition to Doctoral Programs. *CBE Life Sci Edu* **5**: 340–347. doi:10.1187/cbe.06.

Quitadamo, I. J., and Kurtz, M. J. (2007). Learning to Improve: Using Writing to Increase Critical Thinking Performance in General Education Biology. *CBE Life Sci Edu* **6**: 140–154. doi:10.1187/cbe.06.

Wood, W. (2009). Innovations in Teaching Undergraduate Biology and Why We Need Them. *Annu Rev Cell Dev Biol* **25**: 93–112.

Yeong, F. M. (2013a). Encouraging students to take a stand: Using primary literature to support scientific arguments. In *7th International Technology, Education and Development Conference* (pp. 2047–2056). Spain.

Yeong, F. M. (2013b). Using primary-literature-based assessments to highlight connections between sub-topics in cell biology highlight connections between sub-topics in cell. *J NUS Teach Acad* **3**(1): 34–48.

Yeong FM. (2014) Using primary literature in an undergraduate assignment: Demonstrating Connections among cellular processes. *J Biol Edu* Accepted.

Index

Made in the USA
Las Vegas, NV
04 May 2021